Frankenstein's Footsteps

Frankenstein's Footsteps

Science, Genetics and Popular Culture

Jon Turney

Yale University Press
New Haven and London

Set in Garamond by Best-set Typesetter Ltd, Hong Kong
Printed in Great Britain by St Edmundsbury Press

Library of Congress Catalog Card Number 97-80782

ISBN 0-300-07417-4

A catalogue record for this book is available from the British Library.

10 9 8 7 6 5 4 3 2 1

for Dr. D, of course

Contents

Acknowledgements

This book has been a long time coming, so some of the folk I am about to name will doubtless have forgotten all about their contribution. But now is the time to thank the following, for help, ideas, leads, comments, encouragement, or just good conversation, sometime between 1976 and now – Garland Allen, John Durant, Jamie Fleck, Jane Gregory, Jon Harwood, Alan Irwin, Frank James, Everett Mendelsohn, R.S. Rosenberg, Jim Secord, Tom Wilkie, and Jan Witkowski. Thanks to the Wellcome Trust for a fellowship which took me out from behind a features editor's desk, and to all my colleagues in the Department of Science and Technology Studies at University College London, where the book was written. And thanks to students on my course on Popular Responses to Science and Technology, who sharpened my ideas a good deal.

Special thanks to those who read and commented along the way – Brian Balmer, Steven Miller, Dorothy Nelkin, Maurice Riordan, Danielle Turney, and an anonymous reviewer for the publisher.

The text they read would not exist without the help of staff at municipal libraries in London and Manchester, the British Library, especially the newspaper library at Colindale, the University of Manchester, London University, University College London, The Wellcome Institute Library, the Science Fiction Foundation (then) at North East London Polytechnic, the Library of Congress and the Archives Office at Rockefeller University, New York.

My thanks to the editors of *Public Understanding of Science* for permission to use material published in the journal in Chapter 4.

The text was turned into a book by the editorial team at Yale University Press. Thanks to Robert Baldock for asking me to write it, to Candida Brazil for overseeing production, to Beth Humphries for meticulous copy-editing, and to Sheila Lee for picture research.

My largest debt is to Edward Yoxen, who started this one off. But my final thanks go to Eleanor and Catherine, whose keen interest in the progress of Frank's Feet was a spur to finishing it.

The chemical or physical inventor is always a Prometheus. There is no great invention, from fire to flying, which has not been hailed as an insult to some god. But if every physical and chemical invention is a blasphemy, every biological invention is a perversion.

J.B.S. Haldane, Daedalus, 1923

Introduction

> I busied myself to think of a story . . . One which would speak to
> the mysterious fears of our nature, and awaken thrilling horror.
>
> Mary Shelley, 1831

As we take stock, a little anxiously, of our prospects for the new century, one
of the few safe forecasts is that biology will come into its own. You do not
have to be mesmerised by the millennium to see the likelihood of a new technological era, in which we have unprecedented power to manipulate and
control living organisms. Soon, we will fashion other living things, and ultimately ourselves, to suit our own designs. We will engineer bacteria to make
them into factory workers, splice genes into plants to induce them to grow
larger, ripen faster and keep longer, and modify animals to yield more suc-
culent meat or even grow organs fit for human transplant.

In short, we have the techniques to reconstitute living species in ways which
far outstrip the power of traditional selective breeding. And, as our knowledge of the details of many creatures' genetic constitutions increases – as we
map the genomes of yeast, pig or man – we are building the information base
to apply those techniques with precision, to supplant natural selection. We
shall finally combine cultural and biological evolution.

All this amounts to one of the most extraordinary legacies which the twentieth century will bequeath to the twenty-first. The recent history of biology,
from the rediscovery of Mendel's laws at the century's opening, to the solution of the DNA structure just halfway through, to the assembly of the tools
for editing and even rewriting the genetic code in the last thirty years, has
been a series of intellectual triumphs. But they are intellectual triumphs of a

1

particular kind. They are the newest fulfilment of the Enlightenment promise that systematic pursuit of scientific knowledge would lead to 'the effecting of all things possible', in Francis Bacon's famous phrase from the dawn of modern science. As the anthropologist Paul Rabinow writes, the very latest stage in this story, the Human Genome Project, epitomises the co-evolution of science and technology, of understanding and doing: 'the object to be known – the human genome – will be known in such a way that it can be *changed.* This dimension is thoroughly modern; one could even say that it instantiates the definition of modern rationality. Representing and intervening, knowledge and power, understanding and reform, are built in, from the start, as simultaneous goals and means.'[1]

As with modernity in the larger sense, we are not sure what to think about this most modern biology. The great discoveries of the century are justly celebrated in the mass media – in a host of popular books, TV documentaries and magazine and newspaper articles.

And yet, we fear that the triumph will turn sour. The science *is* extraordinary but we are consistently ambivalent about its application. We recognise that there may be great benefits, but also see less well-defined threats in biological technology's ability to break down old boundaries, and dissolve taken-for-granted categories. There is something profoundly unsettling here, something desired but also feared, something which has always been implicit in the Baconian project which becomes ever less appealing the closer we come to achieving it.

In fact, biology's dizzy onward rush from potential to real technology, its final subordination of the natural world to technique, brings to a new pitch tensions which have always existed in our attitudes to modern science. The best evidence for this is that when we look for ways to interpret the latest developments, the hot news from the lab, the technological promises for the twenty-first century, when we look for stories to tell about what we are about to do, we commonly reach back to a story which is almost two hundred years old.

Mary Shelley's *Frankenstein* has long been a versatile frame for interpreting our relationship with technology. But now that we have developed a real biological technology, biology's power over our collective imagination has been reinforced in ways that we urgently need to try and understand. That is why this book, which offers a history of public images of biology, is organised around a discussion of *Frankenstein.* Mary Shelley did not, in any plausible usage, offer predictions about the future of science. But she did, at the very start of the modern era, identify concerns which go to the heart of our response to science. Her story about finding the secret of life became one of the most important myths of modernity. Now that the secrets of life are ours for the taking we need to ask what role that myth will play in the collective debate about how to make use of them.

In this book, I examine the stories we have told about the doings of biologists over the past two centuries. By looking at what stories were most widely heard at various times, it is possible to write an oblique history of the science. That history, to be sure, is related to more conventional histories, of ideas, experiments and institutions. But my account is more concerned with how non-scientists interpreted or made sense of what they thought was happening. My premise is that fictional representations matter, that the science and technology we ultimately see are partly shaped by the images of the work which exist outside the confines of the laboratory report or the scientific paper. If we want to understand the origin of the vocabulary in which present-day debates about science are conducted, we need to attend, not just to the internal development of science, but to the history of science in popular culture. We need to listen to all the people who commented on the science of their day in the spirit of Mary Shelley, when she referred to some experiments of her great contemporary Erasmus Darwin, Charles Darwin's grandfather, with the rider that: 'I speak not of what the Doctor really did, or said he did, but, as more to my purpose, of what was then spoken of as having been done by him.'[2]

I have organised my own story around a discussion of *Frankenstein* to emphasise that it is the governing myth of modern biology. Mary Shelley's story, in all its many manifestations, is the strongest evidence for my principal claim, that there are vital continuities in cultural debate about science over this long period. And I also suggest that these continuities can help us to understand the sources of attitudes to present-day science, at a time when a variety of explanations is on offer for what is seen as opposition to some new science and technology.

The prevalent images of science, particularly life science, are hotly contested. Geneticists, especially, feel that contemporary mass media portray their work in a negative light. My colleague the developmental biologist Professor Lewis Wolpert, chairman of the British Committee on Public Understanding of Science, chastises the media for what he terms 'genetic pornography'.[3] He is one of a number of scientists who claim to see an anti-science movement in modern Western societies. General opinion poll evidence suggests no such thing,[4] but the belief that it exists is still intuitively plausible to many researchers and to those who speak on their behalf. The outgoing editor of *Nature* summed up the feeling in a revealing jeremiad bemoaning 'the prevalent mistrust of science' at the end of 1995.

What is certainly true is that there is *ambivalence* about science and technology, and that this ambivalence is evoked more by some areas of science than others. Modern biologists have noticed this, of course, but they consistently ascribe it to general anti-science sentiment, ignorance or media misrepresentation of their work. The late Sir Peter Medawar put it most memorably in a comment at the time of the debate about genetic manipulation, 'recombinant DNA', research in the 1970s: 'I find it difficult to excuse the lack of confidence which otherwise quite sensible people have in the scientific profession . . . for their fearfulness, laymen have only themselves to blame and their nightmares are a judgment on them for their deep-seated scientific illiteracy.'[5]

Fifteen years later, the long-time professor of bacteriology at Harvard, Bernard Davis, prefaced his collection of essays on *The Genetic Revolution* with the sweeping suggestion that

> public suspicion of genetic engineering not only has had a direct effect on the advance of biotechnology; it has added one more facet to a more general skepticism about the goals and the social impact of science and technology . . .
>
> This antiscience movement poses a threat, more than is generally recognised, to public support of science, the recruitment of promising students, and ultimately the morale of working scientists.[6]

This rhetorical response to perceived public criticism is part of a refusal to engage with specifics, with particular objections to particular lines of inquiry or technological developments. It neglects the possibility that there may be special reasons why an increase in manipulative power in the life sciences might provoke public disquiet. As we become more urgently aware of the need to foster debate about the development and control of the technologies now coming on stream from academic and industrial bioscience laboratories, we must try to find a more satisfying account of what shapes the vocabulary of that debate than the assertion that it stems from 'genetic pornography'.

But how? If we want to make the origins of this vocabulary and imagery an historical question, there are a number of obvious problems. Today, public attitudes to biological science and its uses can be studied in real time, using a range of survey and interview techniques. We can see how complex the pattern of contemporary ambivalences is from the more sophisticated of such surveys. People welcome some developments and are dubious about others, support some lines of inquiry but wish to see others restricted or stopped altogether.[7] But it is a different matter if we want to know what public responses were evoked by specific areas of science in the past.

Whether we look at past or present, we must avoid a simplistic view. There is no single, simple entity called 'science', and neither is there a single 'public'. Both are complex categories, which change over time. Indeed, there are few individuals who have a single, simple attitude or response to science, as the surveys show. Images and attitudes, and their means of propagation, also change over time.[8] We rarely have any direct access to historical publics, so inferences about what they might think about science have to be indirect as well. The way to proceed is to combine as many kinds of sources as possible, while keeping in view the limitations of what each can tell us. The most fruitful sources shift over the period I want to cover. In the nineteenth century, I find the most revealing stories in printed fiction, especially science fiction (supporting those who see *Frankenstein* as inaugurating the genre). At the beginning of this century, press sources are more illuminating. Later on, one must include film as a vital cultural medium. All can contribute to an understanding of the public history of science. In general, I combine a somewhat compressed history of some aspects of experimental biology with a more detailed look at a small number of episodes when there were fairly direct, clear-cut public responses to scientific work – when latent anxieties were made manifest.

My overall aim is to contribute to the cultural history of images, as the American historian of physics Spencer Weart has done brilliantly for nuclear science. As he writes: 'the historian of images can take up a straightforward task: to look through materials of every description and find for a given group of people what pictures, symbols, beliefs, rational concepts, feelings and emotions have become strongly associated with one another in a cluster that includes a particular subject.'[9] This involves paying attention simultaneously to what scientists are actually saying and doing at specified times, and to how they are perceived.

Another way of putting it is to say that one is trying to map social representations of a particular subject, in the sense of the term used by social psychologists.[10] Or one may consider the way a story encodes a 'script' in the sense coined by the neuroscientist Roger Schank. For Schank, scripts of various kinds, from behavioural to emotional, are crucial to the operation of memory, and help us to navigate through a wide range of possible social and cultural encounters. Once a script has been laid down, a single cue can evoke an entire story, as an interpretive frame or context for what is being discussed.[11] In this sense, the *Frankenstein* script has become one of the most important in our culture's discussion of science and technology. To activate it, all you need is the word: *Frankenstein.*

Finally, it is worth asking why we can expect there to be sense in tracing these ideas and images over such a long period: why might there be any continuities over a time of such enormous change in so many areas? Why do we need to read Mary Shelley to understand the importance of Michael Crichton? The episodes I shall consider begin at a time well before the professionalisation of science, or even the coining of the word 'scientist', when biology was many years from being constituted as a separate discipline, and when there was nothing approaching a mass reading public. Yet I want to relate them to debates taking place now, in an age of mass media, big science, and systematic industrial exploitation of biotechnology.

However, even though the world at the turn of the nineteenth century is in some ways scarcely recognisable today, British society at that time did already have some key features which distinguish it from pre-modern societies. The long trajectory which marks out the development of modern societies was already under way. It was already possible to sense what this meant for the experience of modernity.

The American cultural historian Marshall Berman, whose work we shall return to, describes modernity as in the grip of constant change, seen as both a promise and a threat. It is exhilarating and exhausting, tantalising and terrifying. Above all it is open-ended, displacing a medieval world of fixed natural and social order by continual transformation. To be modern, says Berman, 'is to find ourselves in an environment that promises us adventure, power, joy, growth, transformation of ourselves and the world – and, at the same time, that threatens to destroy everything we have, everything we know'.[12]

This is a description of modernity which is easy to recognise today, but the sensation is not new. For some, it began nearly five hundred years ago, according to Berman. Undaunted by the scope of the generalisation, he divides the history of modernity into three phases. The first stirrings occupied roughly the years from the early sixteenth century to the end of the eighteenth. In these times, people in the European countries affected were dimly aware that life was changing, but unable to describe the changes adequately. The second phase, he suggests, begins with the revolutions of the 1790s: 'with the French Revolution and its reverberations, a great modern public abruptly and dramatically comes to life. This public shares the feeling of living in a revolutionary age, an age that generates explosive upheavals in every dimension of personal, social and political life.'[13] The third phase of modernity then comes in the twentieth century, when the modern system embraces the entire globe, and it is marked by greater triumphs and deeper tragedies than ever before.

So to claim that modernity was already taking shape in Mary Shelley's time is more than saying that the industrial revolution had already started in Europe. The most common position taken today by social theorists is that modernity, while hard to define, is best taken to mean a whole complex of associated changes. It refers, as the Cambridge sociologist Anthony Giddens puts it, 'to modes of social life or organisation which emerged in Europe from about the seventeenth century onwards and which subsequently became more or less worldwide in their influence'.[14] This is not the place to discuss the meanings of modernity in detail.[15] The processes indicated include political, economic, social and cultural changes, all interacting with one another. A convenient summary is provided by Giddens' fellow theorist Stuart Hall, who suggests that a modern society is characterised by secular political power expressed through the nation state; a monetarised market economy; decline of a traditional social order and reconstitution of social and sexual divisions of labour; and the rise of a rationalist, materialist culture.[16]

I am neither going to criticise nor defend this particular attempt to reduce this myriad of changes to a simple scheme. My point, in expanding on the idea of modernity in this preliminary sketch, is simply that it suggests a broader historical context for *Frankenstein*. Mary Shelley wrote in exactly the years which Berman sees as marking the end of the first phase of modernity and the beginning of the second. He offers other authors as the key contributors to diagnoses of the modern condition. But Mary Shelley has, I think, left as lasting a mark on the debate about the meaning of modernity as anyone.

This is why we find ourselves searching for the sources of such persistent imagery representing science at the start of the modern period. I will take up the response to modernity again, but for now let me simply suggest that *Frankenstein* is best read as a myth of modernity. And it can now be seen as such because of the turn modernity is finally taking, two hundred years on.

To understand the importance of this myth, we need to know something of the origins of Mary Shelley's tale. This is discussed in Chapter 1, which offers a sketch of the milieu from which *Frankenstein* emerged, considers the varied sources – literary, autobiographical and scientific – for Shelley's story, and gives an account of the many and varied critical interpretations of the novel.

The next chapter moves on to the after-life of the story, and the myriad retellings of the tale of an obsessive scientist creating life. This is the place for an explanation of the idea that the story deserves to be recognised as a modern myth, and for a summary of the essential elements of the myth: the features of Mary Shelley's complex narrative which have survived all the retellings to become part of the *Frankenstein* script. Mary Shelley, I argue, produced a story which expresses many of the deepest fears and desires about modernity, especially about violation of the body. The human body is both a stable ground for experience in a time of unprecedentedly rapid change and a fragile, limited vessel which we yearn to remake.

In Chapter 3 I turn to other nineteenth-century stories, after a very brief review of the rise of experimental biology in the years after *Frankenstein*. A feature of this newly professionalised discipline was the use of living animals in the laboratory, and stories told to a greatly enlarged reading public by the antivivisectionists of the 1870s find echoes in later Victorian tales, from the years of two enormously powerful popular fictions, Robert Louis Stevenson's *Dr Jekyll and Mr Hyde*, and, just before the turn of the century, H.G. Wells's *The Island of Dr Moreau*. Although non-scientists were eager to share in the medical benefits of the new experimental biology, their ambivalence can be seen in stories like *Moreau*, which explores the dark face of science in a detail that earlier writers had never achieved.

Moving into the twentieth century, Chapter 4 begins a detailed analysis of a remarkable period in the public history of biology, when press and public really did believe that scientists were on the verge of creating life in the laboratory. They believed it largely because of the bold claims made by some prominent biologists of the time, most importantly by the great propagandist for a reductionist research programme, Jacques Loeb and his colleague at the new Rockefeller Institute for Medical Research in New York, the transplant surgery pioneer Alexis Carrel. Both courted notoriety, courtesy of a press hungry for news from the laboratory, and became two of the earliest 'visible scientists' in the United States. Responses to their work, at a time when optimism about the prospects for continued scientific, technological and social progress was nearly universally shared, were nevertheless deeply ambivalent.

The chapter finishes by following the same story, but in Britain, where another celebrated experimental biologist, Edward Schafer, presented Loeb and Carrel's claims to the nation in a famous presidential address to the British Association in 1912. The extraordinary volume of press coverage of Schafer's speech, in which references to *Frankenstein* abounded, attests to a similar ambivalence about the prospect of biological power to that found on the other side of the Atlantic.

The story moves on into the twentieth century in Chapter 5, taking in the ground-breaking discussions of biology and society of the 1920s by such figures as Bertrand Russell, J.B.S. Haldane and Julian Huxley. Their constant refrain – that the physical sciences were as nothing compared with the sciences of life when it came to changing the conditions of life – was explored in the next great work of fiction to enter the vocabulary of popular debate, *Brave New World*, by Julian Huxley's brother Aldous. Although his main idea, of uniting life creation with modern mass production, had been prefigured in the 1920s by the Czech writer Karel Čapek in his play *R.U.R.*, it was *Brave New World*, drawing on Haldane's ideas, which established the test-tube baby as a potent symbol of the project of twentieth-century biology.

In Chapters 6 and 7 I review the extraordinary development of biology between the 1930s and the 1960s. Even though the advent of nuclear weapons underlined the powers that could be acquired through applied physics, the perception that biology was the true science of human transformation persisted. And a biology which finally appeared capable of bringing this about was being fashioned in the laboratories now lavishly endowed by the state. An early comment on this capability was the French biologist Jean Rostand's popular book *Can Man Be Modified?* In the 1960s, the representation of molecular biology as a discipline which had fathomed the 'secret of life' – the structure of DNA – together with a whole set of new technologies in surgery, contraception and pharmacology, was used to support the idea that we were witnessing the first stirrings of a 'biological revolution'. Chapter 7 analyses how this notion was constructed, and considers the early responses to it in books like Gordon Rattray Taylor's *The Biological Time Bomb*, a return to Rostand's arguments from the 1950s augmented by reports of the new biology of the following decade.

Two episodes from the 1970s brought the potential of the biological revolution home to its various publics: the birth of the first 'test-tube' baby in 1978 and the debate over recombinant DNA research which began in the mid-1970s. The following two chapters each considers one of these episodes. Chapter 8 reviews the discussion of *in vitro* fertilisation, one of the mainstays of the emerging discipline of bioethics. I show how often the leading writers on bioethics of the time drew on Huxley's images to dramatise their arguments, and how the same images recurred in reporting of the eventual birth of Louise Brown in Britain. This event evoked the now familiar ambivalence about where biology might be heading, offering a widely celebrated result but a celebration always accompanied by unease about the kind of reproductive intervention which made it possible.

Chapter 9 reviews the debate about recombinant DNA research, using the work of other historians to argue that there was a 'hidden agenda' to the debate, in the US and Britain. While most of the discussion focused on containing possible health or environmental hazards of manipulating DNA in the laboratory, there were wider concerns about the work which remained largely untouched. *Frankenstein* and *The Andromeda Strain* coloured people's views about what the work might mean, but were rarely acknowledged as presences in the discussion.

What role do we then find *Frankenstein* and kindred fictions playing in public deliberations about biological research in the 1990s, the era of the Human Genome Project? The final chapter examines the state of public images of experimental biology today, showing how potent a force the old scripts still are. They can be seen at work in films like *Jurassic Park* or, needless to say, Kenneth Branagh's *Mary Shelley's Frankenstein*. A host of other literary and broadcast productions suggest that our ambivalence about power over the body is as pronounced as ever. But the 1990s are also seeing a keener awareness on the part of scientists that these images matter, and that they can be contested. Indeed, it appears that our common knowledge of the *Frankenstein* myth means that it can be a rhetorical resource for proponents of the latest research possibilities as well as their critics – as was evident in the British debate of the 1980s about experiments on laboratory-fertilised human embryos.

In the end, though, it is a resource which, while it is bound to continue to find uses in such debates, hampers the necessary task of agreeing how to control the new technological powers now being developed in the laboratories. It invites an all-or-nothing response to a whole complex of developments, when we should be insisting on our right to choose some, and block others. When we do so, it should be for reasons which we can articulate more clearly than saying either that there are some things humans are not meant to know, or that we should not tamper with nature. But to prepare to articulate those reasons, first we must come to terms with *Frankenstein*.

I offer here, then, a consideration of stories about biologists, and I stress throughout that the development of ideas in and about the life sciences is one important way in which we have worked out many of our feelings about the modern project. I review a very wide range of commentaries on the possibilities of biological manipulation, and try and show how they have developed through changing times – as public debate, media and science have grown and changed.

I argue both that attitudes to biology are a key to the *Frankenstein* myth, and vice versa. While *Frankenstein* has common use as a symbol for technology in general, I want to focus on its specifically biological aspects. This leads me to say rather little about another thread in the history of science and technology which often evokes Frankensteinian images. The story of the mechanical human is one which I touch on at many points here, but do not take up in any detail. I have not tried to incorporate here much of the huge volume of comment on artificial people made by non-organic means, about robots, computers or artificial intelligence. In particular, I almost completely avoid mention of two currently popular symbols of the potential of new technologies – the computer programs which simulate natural selection known as A-life, and the human-machine hybrid, or cyborg.

All these stories can tell us much about attitudes to technology, and our anxieties about the boundary between human and non-human. But they are almost all, in my judgement, still mostly concerned with fantasy, not reality. The near-obsession with the idea of the cyborg in much writing on postmodernity, which builds on a preoccupation with such figures in recent Hollywood films, is a fashion with rather little technological warrant. It is true that we have fancier prosthetics now than hitherto, but some commentators seem to think that the possession of false teeth makes one a cyborg.

What the cyborg literature does tell us is that the impulse to imagine a future in which we transcend the body is as strong as ever. This is a powerful theme in twentieth-century speculations about technology, most memorably explored in J.D. Bernal's *The World, the Flesh and the Devil* published in the 1920s. The idea that we can become one with our machines, immortal, invulnerable, or even become disembodied and live in a computer-generated virtual reality, is as compelling a focus for our ambivalence about technology as the notion of creating or modifying life by artificial means.

I do not say that this debate does not matter because none of the technology required is yet with us. Imaginings of this kind plainly shape real technological agendas over long time-spans.[17] However, the technology to manipulate life is much nearer yielding the kind of power over its objects of study which speculation envisages than the cyborg project. And, despite a blurring of boundaries in some respects, there is still a distinction worth making between the organic and the non-organic realm. In addition, I would argue that there is a stronger sense of a real discontinuity in the current realisation of an ability to alter life forms than in the fashioning of ever more sophisticated mechanical or electronic aids. That can be seen as a straightforward continuation of an old story.[18] Another way to put it is to suggest that applied biology threatens a different set of boundaries between categories. Show me someone with a heart pacemaker, and I have no real difficulty in seeing which part is human, which is machine. Show me a ewe whose genes have been altered so that it secretes a human protein in its milk, and it is much less clear which part is human, which sheep.

In any case, whether one sees either biotechnology or electromechanical technology as the harbinger of qualitative rather than quantitative change in the human condition, the cyborg and the robot are part of a story which can be separated from the one I want to tell here – and which is being explored increasingly often by other writers.[19] This book is about Frankenstein the biologist.

Chapter One

Mary Shelley's Creation

Inside Mary Shelley's novel lie the seeds of all later diseased creation myths.

Brian Aldiss, 1975, p. 29

Promethean and alchemical origins

What happens in *Frankenstein*? The question readily evokes a simple enough script. A man, a scientist, creates a living being which, grown monstrous, turns on its creator. Into this general frame fit a thousand stories. Creator and created take many forms, and the details of the consequences may vary, but these are the essentials.

It still comes as a surprise to new readers that the book from which this popular cultural script ultimately derives is an altogether more subtle and complex creation. In the next chapter, I will outline the transformations of the story at the hands of playwrights, writers, film-makers and comic book artists which have made it such a pervasive cultural presence. But I want to start at the point of origin, with the novel to which these other artists repeatedly return, and to explain what distinguishes it from earlier myths.

There are many themes in the novel *Frankenstein*, but the theme of the *Frankenstein* myth is the getting and using of knowledge, and the power that knowledge may confer, a power dramatised by the creation of life. Ambivalence about knowledge is, of course, a staple of myths in many cultures, from Prometheus to the Garden of Eden.[1] So what is it about *Frankenstein* which makes it distinctive, so distinctive that it has become the standard script for expressing this ambivalence? And what was it about Mary

Shelley and her circle which enabled her, a gifted 18-year-old making her first serious stab at writing, to create such a story? *Frankenstein* has attracted an enormous amount of critical attention over the last thirty years, giving us many possible answers to these questions. In this chapter I summarise the main features of the original text, as well as discussing how it was composed. This account of the genesis of the book, the moment of Frankenstein, is followed by a look at its growth and transformation into the myth of the later years, with its myriad retellings across all media. First, though, we should look at the young Mary Shelley's novel in the context of earlier literary traditions.

Frankenstein marks a transition, in stories of men creating life, because Victor does not invoke the aid of the Deity, or any other supernatural agency. He achieves his goal by dint of his own (scientific) efforts. Earlier myths of life creation fall into two main categories, each of which touches on a fundamental human concern. In the first, the searcher for the secret of life wants to probe the origins of mankind in the Creation – the gods must be tricked into performing the first and greatest miracle a second time, or into disclosing the means to a mere mortal. The second, related, theme is how knowledge of the wellsprings of life can help someone evade the finality of death, regaining a lost immortality or gaining assurance of reincarnation. Variations on these themes may be found in the myths of all cultures.[2] These stories very often fall into the larger class of myths dramatising ambivalence about knowledge, and about the power that possession of knowledge confers. Myths relating the acquisition of knowledge almost always end badly, and when the knowledge obtained gives the power to create life or postpone death, the consequences are dire indeed. Think only of Prometheus chained to his rock by a vengeful god, his liver daily devoured by a predatory bird, only to regenerate overnight and be devoured anew. Although Prometheus' original tormenter was Zeus, the Promethean legend became transformed in Roman times into an enduring symbol of the consequences of mocking the Christian God.

The second principal Western myth of a man-made being is the golem of Judaic tradition. As in the Promethean legends, a clay figure's life is God-given. And like Prometheus, the golem is also a clear precursor of Mary Shelley's creature, being gigantic, fearsomely difficult to control, and often functioning as a *doppelgänger*, embodying the evil in the soul of its creator. In the end, creator and created destroy one another, though in many versions the golem has first saved a Jewish community from attacks by outsiders.

The gradual emergence of a materialist philosophy made it possible to imagine a larger direct contribution to life creation by earthbound protagonists. During the Middle Ages, some tales turned from men who fashioned a crude simulacrum, and infused life into the figure with supernatural aid, to those who actually fabricated their creatures according to designs derived from their studies of nature. Many stories were woven around the supposed achievements of such historical figures as Albertus Magnus in the thirteenth century, who was reputed to have constructed a servant from brass, the alchemist Cornelius Agrippa, and Paracelsus who left among his voluminous writings of the sixteenth century a suggestion about how to generate a homunculus from blood, faeces and semen.[3]

Alternatively, one might build an automaton. With improvements in the mechanical arts, the construction of more and more impressive automata became feasible, and they provided an intense focus for the philosophical debates of the Renaissance. Cartesian mechanism essentially portrays all organisms as automata. In Descartes' physiological system, set out in his *Treatise on Man*, the difference between a living and a dead body was analogous to that between a watch wound and unwound. Of all living things, only man possessed a soul. Descartes' presentation of these ideas was tentative. As he was about to publish his treatise on *The World*, of which the *Treatise on Man* forms the second part, Galileo was condemned by the Inquisition in Rome, and he vowed never to release it. On its eventual publication after his death the book assured readers that Descartes' subject was not man but a machine, and not a real but a hypothetical machine such as God might have made had he wanted the result to resemble a man. It was, in Thomas Hall's phrase, an 'ambiguous robot'.[4]

Nevertheless, although he drew the line at identifying man as a machine, the implication that all the essential features of a man, bar the soul, could be duplicated by a sufficiently artful mechanic was resisted. In a satire directed against the Cartesian world picture, Spinoza wrote that the golem 'has as much life as any human being, if one accepts that the relation between body and mind is so loose that it can in a moment be lifted and replaced'.[5]

Descartes' disciple La Mettrie carried the idea of creating life from matter one stage further. In his *L'Homme machine* (1747) he denied the distinction between animate and inanimate matter, and between body and soul. In La Mettrie's monism all the properties of life were immanent in ordinary matter, but only emerge in the correct form of organisation. The soul became a property of the organisation of the matter of the brain.[6] In other words, in this scheme there was no bar to fashioning animate from inanimate matter. An oft-quoted passage shows the influence of the crowd-pulling automata of the period on La Mettrie's conception of life, as well as explicitly linking his extension of Cartesian mechanism to the Promethean legend:

> Man is . . . to the ape . . . what Huyghen's planetary clock is to a repeater watch made by Julien le Roi. If more gadgets, more cogwheels, more springs were needed to register the movements of the planets than to register the hours, or repeat them; if Vaucanson needed more ingenuity to construct his *flute-player* than for his duck, he would have needed still more to construct a *talker* – a machine which can no longer be dismissed as impossible, particularly in the hands of a new Prometheus.[7]

The last sentence, of course, anticipates the subtitle of *Frankenstein*, which draws a great many elements from the myths and stories which preceded it, as well as incorporating the possibilities of the newly developing mechanistic view of living things.[8] But in Mary Shelley's novel, which appeared seventy years after *L'Homme machine*, the creator of life was for the first time recognisable as a scientist. How exactly did this come about?

A little literary competition

Mary Wollstonecraft Shelley's *Frankenstein, or The Modern Prometheus* was first published in an edition of three volumes in 1818 in London. The story has its own origin myth, based on the account Mary later gave of the genesis of her tale, which is still retold in introductions to contemporary editions of the novel and has attracted its own film and television treatments. Mary had been struggling to think of a ghost story to entertain her companions during a rainy summer spent by Lake Geneva. She was in the company of two of the most brilliant young literary men of her time, her lover Percy Shelley and the equally charismatic Byron. They were each to try and produce a tale to match a volume of ghost stories they had been reading, 'One which would speak to the mysterious fears of our nature and awaken thrilling horror.'[9] The 18-year-old Mary, yet to make her mark as a writer, found it hard to get her creation under way. Thus preoccupied, as she described in a famous passage in the 1831 preface, she had a waking dream, a habit she fell into in childhood. This time, her vision left her deeply unsettled: 'I saw the pale student of the unhallowed arts kneeling beside the thing he had put together. I saw the hideous phantasm of a man stretched out, and then, on the working of some powerful engine, show signs of life, and stir with an uneasy, half-vital motion.'[10] If true, her dream did indeed give her the scene at the heart of the novel. It seemed to her the obvious subject for her ghost story; 'What terrified me will terrify others.' The dream-scene was quickly turned into the most enduring image of the book. The words that open Chapter 4 of Volume I were also the first words of her original short story:

> It was on a dreary night of November that I beheld the accomplishment of my toils. With an anxiety that amounted almost to agony, I collected the instruments of life around me, that I might infuse a spark of being into the lifeless thing that lay at my feet. It was already one in the morning; the rain pattered dismally against the panes, and my candle was nearly burnt out, when, by the glimmer of the half-extinguished light, I saw the dull yellow eye of the creature open; it breathed hard, and a convulsive motion agitated its limbs.[11]

That image, surely, is part of our common culture. But there is much more to the novel which Mary worked up from her ghost story, and much more to its creation. In the novel, the story begins with letters written by Robert Walton to his sister back in England. Walton is an explorer, captain of a ship carrying an expedition he hopes will reach the North Pole. Walton, the friend-less leader of an unwilling crew, is himself an obsessive knowledge-seeker. As he later tells Victor Frankenstein, at the pole he hopes to

> discover the wondrous power which attracts the needle; and may regulate a thousand celestial observations, that require only this voyage to render their seeming eccentricities consistent for ever. I shall satiate my ardent curiosity with the sight of a part of the world never before visited, and may tread a land never before imprinted by the foot of man. These are my enticements . . . (p. 7)[12]

As they try to traverse the ice-bound seas, Walton's crew spot a sledge on a passing berg, and thus rescue Victor. Brought aboard, he faints, and is revivified in the traditional way: 'we restored him to animation by rubbing him with brandy, and forcing him to swallow a small quantity' (p. 14). He recovers enough to tell Walton his story, recorded in the captain's journal. This becomes the second level of what is to be a three-layered narrative. Here we are on much more familiar ground. The young medical student, Victor Frankenstein, leaves his native Geneva for the University of Ingolstadt. There, he is seized by the obsession that he can uncover the secret of life. The test is to animate a creature which he builds from the parts of dismembered corpses. When he succeeds, he flees in horror from his creation, and spends a summer trying to forget its existence. During that summer, Victor's young brother William is found murdered, and circumstantial evidence convicts a family servant, Justine, who is hanged for the crime.

The real murderer is, of course, the creature, whose self-education forms almost all of Volume II. He tells his story to Victor, in the third layer of the narrative, when they meet again on a frozen mountainside. He is an eloquent, passionate being, familiar with all of Mary Shelley's favourite works of literature, but with a burning animus against Victor, the father who abandoned him at birth, and against all the other naturally born humans who are repelled by his appearance. Victor, appalled, pitying, guilty, frightened and threatened, agrees to create a second, female, being as a companion for the creature. He embarks on this new project, but destroys the female before it is complete, his conscience finally awakened by the prospect of his new Adam and Eve uniting to produce an entire race of their like. There ensues the creature's bitter revenge: first he murders Victor's fiancée on their wedding night, then he leads a despairing Victor on a deadly chase to the far north. Victor dies on board Walton's ship, and the creature makes a final appearance, both to gloat and to mourn. He declares that he has nothing further to live for, will journey to the pole himself, and build his own funeral pyre there. The being, as Walton calls it, springs from the ship and is soon 'lost in darkness and distance'.

Thus, in bare outline, the story runs, but this is too bald a summary of this new creation myth. For Mary Shelley's complex, youthful text is an imperfect synthesis of many influences, influences which have made it, in Marilyn Butler's phrase, famously reinterpretable. To piece some of these together, we need to go more deeply into the background. Here one can draw on the wealth of recent *Frankenstein* scholarship, which includes material analysing the text from the viewpoints of every school of literary analysis,[13] as well as, for instance, Anne Mellor's excellent scholarly biography of Mary Shelley, which also discusses the novel at length.[14]

The influences all this work discloses are many. The book is indelibly marked by Mary Shelley's turbulent life: she was the daughter of two great intellectual radicals, William Godwin and Mary Wollstonecraft, the mother who died immediately after her birth; she was Shelley's teenage lover (his wife before the book was finished), and already the mother of a premature baby daughter of her own, who lived less than two weeks, and an infant son. There were other estrangements, and other deaths, along the way, and to come. It all added up to 'the elaborate, Gothic psychodrama of her family history', which is deeply inscribed in the novel.[15]

It is marked, equally clearly, by the turbulence of the era. Mary grew up at a time when it seemed that the French Revolution might be followed by the overturning of the established order in Britain, at least until the Reform Act of 1832. She belonged to a society seeing the first real effects of industrialisation, when whole landscapes marked by 'dark satanic mills' were becoming visible.[16] And she was witness to the growing power of science to demonstrate experimental control of natural phenomena, in the study of electricity by Galvani and Volta, for example, or the new chemistry of Priestley, Davy and Dalton.

The book which Mary Shelley composed was a remarkable blend of literary and scientific sources, made possible by her unusual education by her father, her later encounters with contemporary science, and her highly charged personal experience of family life and reproduction. By the time she wrote her first novel, Mary knew more about sex and death, as well as the life of the mind, than women of more conventional background. There is no doubt that the tale was conceived during that summer of 1816 which Mary and Percy Bysshe Shelley, not yet married, spent in Switzerland, in the company of Byron and Dr Polidori, Shelley's secretary and an Edinburgh-trained physician. The interests of each played some role in the final shape of the tale. The proposal to write a ghost story was Byron's, and Mary's was the only one to be completed at the time.[17] This inspiration was wedded to ideas and suggestions from a wide variety of sources to produce a tale which is a precarious balance between old and new; a Gothic novel with a scientific rationale for fantastic events; a Faustian legend where Faust is spared divine retribution but cannot himself bear the responsibility for his creation.

Mary might have been expected to turn to alchemy or mysticism for her animating principle. Her father, William Godwin, wrote a *Lives of the Necromancers* which included chapters on Paracelsus and Agrippa, and it is indeed the writings of these figures that first seize the imagination of the young Victor Frankenstein. However, when he progresses from self-directed study to the university, the professors scoff at his infatuation with Albertus Magnus and Paracelsus, and he becomes acquainted with modern chemistry.

In Brian Aldiss's words, 'symbolically, Frankenstein turns away from alchemy and the past towards science and the future – and is rewarded with his horrible success'.[18]

The prominence of science, and its relation to older mythical motifs, was a major preoccupation of Byron and Shelley in their exhilarating late-night conversations. Both poets were gripped by the Promethean theme. Shelley, in particular, was fascinated by automata, and by the new chemistry. He was also much influenced by Erasmus Darwin, in both his poetic style and his scientific outlook. Darwin, in fact, became one important source for the idea of life creation by the aid of science, at least according to the prefaces which the two Shelleys wrote for successive editions of the novel. In the original preface, Percy Shelley, writing anonymously, asserted that 'the event on which this fiction is founded has been supposed by Dr Darwin and some of the physiological writers of Germany as not of impossible occurrence'. Similarly, as I noted in the introduction, Mary Shelley wrote in 1832 of listening to Byron and Shelley's talk of the experiments of Dr Darwin.[19]

Her readers would have known very well who she meant. Erasmus Darwin was a man of prodigious talents and radical outlook. The leading physician of his time, he shared with Shelley his atheism, materialism, commitment to free thinking and interest in technology, botany and medicine. His poetry, famous in his lifetime, was an important formative influence on Shelley's own style.[20] He partook of all the great scientific discussions of the late eighteenth century, and was forever engaged in speculating about their results. In consequence, just as he prefigured his grandson's formulation of a theory of organic evolution, he also anticipated the biologists of a century later by his suggestion that life might one day be created. In his major works, *Zoonomia*, published in 1794, and *The Temple of Nature*, he included both his evolutionary speculations and hints of the creation of life by man. These eventually resurfaced in Byron's and Shelley's conversation as the suggestion that Darwin had preserved a piece of vermicelli in a glass case, 'till by some extraordinary means it began to move with voluntary motion'. Mary's tale thus had at least some warrant in the natural-historical speculation of the time, as she and Percy Shelley were at pains to point out, even though in her later formulation she distanced herself from the claim that Darwin himself had actually said life creation might be possible.

How significant was such a claim? Critics dispute the importance of the scientific elements in *Frankenstein*. Brian Aldiss, in his influential history of science fiction, takes Mary Shelley's story as the first work which embodies all the essential characteristics of the genre, serving both to inaugurate and to define it, a judgement which others have endorsed.[21] On the other hand, the critic James Reiger argues that it is a mistake to call *Frankenstein* a pioneer work of science fiction: 'Its author knew something of . . . Davy, Darwin and Galvani . . . but Frankenstein's chemistry is switched-on magic, souped-up alchemy, the electrification of Paracelsus . . . He is a criminal magician who uses up to date tools.'[22]

This, however, is a crucial innovation. Mary Shelley's creature belongs to a different age and a different set of beliefs about the universe to the homunculus.[23] It is precisely the electrification of Paracelsus which marks out Frankenstein as a pivotal point in the transition from the supernaturally fantastic to the scientifically plausible. The chemist Waldman at the university tells Victor how the ancient teachers 'promised impossibilities and performed nothing'. On the other hand, modern scientists

penetrate into the recesses of nature, and show how she works in her hiding places. They ascend to the heavens; they have discovered how the blood circulates, and the nature of the air we breathe. They have acquired new and almost unlimited powers; they can command the thunders of heaven, mimic the earthquake, and even mock the invisible world with its own shadows. (p. 28)

The passage is a careful pastiche of Humphry Davy's contemporary chemical rhetoric, although echoes of Francis Bacon can also be heard clearly. We do not know if Mary Shelley met Davy but he was known to her father: he was one of the many who dined at the political philosopher's table. And Percy Shelley's youthful enthusiasm for chemistry had stayed with him to the extent that he urged Mary to read Davy's *Elements of Chemical Philosophy*. In her novel Mary implies that the science which Davy saw yielding such great power over the natural world will ultimately realise the ancient dreams of control over life and death.

Although the precise details of the creature's animation are carefully obscured in the novel, there are hints that electricity is involved. For contemporary readers this undoubtedly lent plausibility to the idea that life might be bestowed on dead flesh. Percy Shelley had lessons on natural philosophy at Eton from Adam Walker, a well-known popular lecturer, who argued that there was a connection between electricity and life.[24] The association was reinforced in the popular mind by public demonstrations, widely described in the press at the time, in which electrical stimulation was applied to the bodies of recently executed criminals.

Electrical provocation of corpses was a tradition established by Giovanni Aldini, Galvani's nephew, in 1804. Contemporary accounts suggested that during these experiments, 'the body became violently agitated and even raised itself as if about to walk, the arms alternately rose and fell and the forearm was made to hold a weight of several pounds, while the fists clenched and beat violently the table upon which the body lay'.[25] This kind of 'reanimation' became the centrepiece of theatrical demonstrations by Aldini and others, including famous occasions in London and Edinburgh. They were a potent symbol of the scope of the new science, although subject to the same kind of reservation as would attend publication of the novel. The *Edinburgh Review* noted that Aldini's experiments were 'rather more disgusting than instructive'.[26]

Altogether, the text of *Frankenstein* suggests that Mary Shelley was extremely well acquainted with the science of her time.[27] But there is a further layer of scientific associations to emphasise. For aside from these connections with scientific ideas and controversies, *Frankenstein* also makes much use of other aspects of science and scientific attitudes, especially attitudes to the body. The provision of bodies for anatomical instruction, and the habit of grave-robbing, brought images from the Gothic novel into the heart of sophisticated late eighteenth-century life. Mary Shelley's slightly later tale may also be read in the context of contemporary debates about anatomy, dissection and grave-robbing.[28]

The dissection of executed criminals in the cause of medical education was a focus for ambivalence about medical knowledge well before the 1818 edition of the novel, and popular unease was exacerbated by the Murder Act of 1752. So Mary Shelley may well have been deliberately drawing on an awareness of this 'legitimate' trade in corpses, as well as of the regular disappearance of corpses from burial grounds – more than a thousand a year between 1800 and 1810. Soon after *Frankenstein* appeared, popular suspicion that grave-robbing was not the only source of cadavers was confirmed by the conviction of Burke and Hare in Edinburgh for murdering indigents in order to sell their bodies to a local surgeon, Robert Knox. But even before the novel was written it was already well established that the most conspicuous contributors to advances in anatomy were dissectors of human cadavers on a heroic scale. Bichat, for example, reported in his *Recherches physiologiques sur la vie et sur la mort*, published in 1800, how he had used six hundred corpses in six months. Many came fresh from the guillotine.[29]

Bichat, in turn, was the bearer of a tradition first laid down by Vesalius and his followers in the sixteenth century – the tradition of learning about the body by taking it to pieces. The cultural historian Jonathan Sawday has described in compelling detail how the early modern period was marked by the emergence of a 'culture of dissection', a mode of inquiry which transformed understanding of the body. He emphasises 'the fear (or desire) which the prospect of anatomical knowledge of the body's interior seems to have excited'.[30]

Mary Shelley drew on this current of fear and desire when she described Victor's descent into graveyards and charnel houses in pursuit of his rather more advanced anatomical studies: 'To examine the causes of life, we must first have recourse to death. I became acquainted with the science of anatomy: but this was not sufficient; I must also observe the natural decay and corruption of the human body' (p. 30). As his studies progressed, Victor recalled, 'I collected bones from charnel houses; and disturbed, with profane fingers, the tremendous secrets of the human frame' (p. 32).

So Mary Shelley's original text includes, among other elements, a creation myth based on science as a substitute for God, a surprisingly realistic composite picture of contemporary science, and a refracted image of the dark side of scientific, and especially medical, practice of the time. It is possible to track down the precise position she takes on contemporary scientific issues.[31] But suffice to say that a view of science and its possibilities, depicted as control over the fundamental biological realities of life, lies at the heart of the novel. It is about a man discovering the secret of life for himself:

> I paused, examining and analysing all the minutiae of causation, as exemplified in the change from life to death, and death to life, until from the midst of this darkness a sudden light broke in upon me – a light so brilliant and wondrous, yet so simple, that while I became dizzy with the immensity of the prospect which it illustrated, I was surprised that among so many men of genius, who had directed their inquiries towards the same science, that I alone should be reserved to discover so astonishing a secret. (p. 30)

My emphasis on the novel as a response to the powers of science will be over-literal for some tastes. This interpretation has to contend with numerous others, all of which can claim some warrant in the text. In literary terms, as well as the conventions of the ailing Gothic tradition, the writing draws on *Faust*, *Paradise Lost*, the 'Ancient Mariner' and on William Godwin's own novels. The book's concerns with birth, sex and death are inseparable from the vision of life creation at its heart, but it plainly uses many other sources, historical and psychological. It is a tale of abandonment, of the search for a family, of the corruption of a Rousseauesque noble savage by a cruel world, of failed parenting, of frustrated sexuality, of reproductive agony and of birth as a harbinger of death. It is an expression of Mary's ambivalence about her relationship with her own father, with Shelley – aspects of whose character appear in both Victor and his creature, and Victor's great friend Clerval – and about her own motherhood. The child William who dies at the creature's hands is the twin of her own son William, her first child to survive into infancy. And there are other readings still – of *Frankenstein* as a text about writing and reading, as a reflection on imperialism; of the monster as the proletariat, or even as the embodiment of a morally superior vegetarianism.[32]

More often in recent years, and closer to my own theme, the novel has been reinterpreted as a feminist critique of science. Mellor argues persuasively that the second principal novelty of the story is Victor's personification of science as a male enterprise, bent on dominating a feminised nature.[33] Victor creates a new life without female aid, destroys the creature's potential mate in a scene figured as a virtual rape, and brings about the death of his own partner before the marriage is consummated. Here, Mary Shelley's own ambivalence about birth, parenting and marriage is interwoven with her attitude to the new science, much as present-day reactions to reproductive technology mobilise the most personal feelings in response to each new laboratory triumph. Men cannot bear children, but can women bear motherhood? Perhaps both might find salvation if we could create life in the laboratory.

But this is to anticipate too much at this stage in my story. Before considering which of the many possible interpretations of the book to adopt, we must consider the fact that *Frankenstein* is much more than just a striking early nineteenth-century novel. All of this scholarly work focuses mainly on Mary's text, but it would not have been pursued so assiduously if that text had not taken on a life of its own in a way which is almost without parallel in modern culture. The critical game is endlessly diverting, and occasionally illuminating. But it begins to produce diminishing returns as more and more effort focuses on the original text, without much heed being paid to the empirical content of the huge volume of succeeding versions of the story. As Brian Aldiss says:

For a thousand people familiar with the story of Victor creating his monster from cadaver spares and endowing them with new life, only to shrink back in horror from his own creation, not one will have read Mary Shelley's original novel. This suggests something of the power of infiltration of this first great myth of the industrial age.[34]

What, then, became of Mary's tale? And what does it mean to call *Frankenstein* a modern myth?

Chapter Two

Hideous Progeny: *Frankenstein* Retold

The truth of a myth . . . is not to be established by authorising its earliest versions, but by considering all its versions. The vitality of myths lies precisely in their capacity for change, their adaptability and openness to new combinations of meaning. That series of adaptations, allusions, accretions, analogues, parodies and plain misreadings which follows up on Mary Shelley's novel is not just a supplementary component of the myth; it *is* the myth.

Baldick, 1987, p. 4

A subterranean and invisible diffusion

Mary Shelley's novel is remarkable in many ways, but the most remarkable thing about it is the way the story has become embedded in our culture. The bulk of the textual criticism naturally focuses on the novel: but critics are aware that there is something out of the ordinary about *this* text. The story has had a life beyond the original book, evolving in ways which are hard to pin down exactly, although they are equally hard not to notice. When the critics try to express what those properties are, the words that appear most often are drawn from a pre-literate world: folklore, legend or myth.

The first definition of myth given in the *Oxford English Dictionary* is 'a purely fictitious narrative usually involving supernatural persons, actions, or events, and embodying some popular idea concerning natural or historical phenomena'. This gives us a place to start in considering in what sense *Frankenstein* may have become a modern myth, when other stories of the time did not. For there is more to this than the casual observer might suppose. There are Dickens novels which everyone 'knows' even though they have not read the original. Costume drama treatments of Jane Austen have been a recent staple of British television, discussed in a hundred newspaper reviews and round countless dinner tables. Yet it would sound odd to write of the 'myth' of *Oliver Twist* (1837), or of *Pride and Prejudice* (1813). The term denotes something both more pervasive, and more elusive. There *are* other examples, *Dracula* being the most obvious comparison with *Frankenstein*, but they are rather few.[1] They achieve a recognisability which is not attached to any particular version or medium. As Christopher Small suggested in his study of the Shelleys, 'mythology is metaphorical thinking in which the metaphor assumes independent and continuing existence', which is akin to Bauer's notion of a social representation taking on a life of its own.[2] In this light, says Small, 'it might be argued that the myth has had most hold where the book hasn't been read'.

That paradox means that some literary critics, while identifying *Frankenstein* as a myth, are a little uneasy about it.[3] Chris Baldick starts one of the most interesting treatments of the idea with the confession that a modern myth is a critical anomaly, even a scandal.[4] What he means is that myths, in the more elaborate anthropological sense analysed by Lévi-Strauss, are seen as elements of a pre-literate culture. A myth is flexible, precisely because it is a characteristically oral form, but retains a stable core of meaning. A printed literary text, on the other hand, is fixed in form, and the critics' job is to discover its many possible meanings.[5]

Baldick argues for a broad interpretation of the myth, embracing human relations as well as technological development, and supports this through a reading of a range of key nineteenth-century novels. But he says little about the broader popular culture. In his view, 'because myths exist in a latent state of subterranean and invisible diffusion in the cultures which adopt them, it is possible after the event to catch them only at those moments when they have surfaced in some documentary form'.[6] This is undeniable, but the documentary forms which are recoverable are much more diverse than the novels he treats in his study. Just how diverse can be seen by inspecting Donald Glut's *The Frankenstein Catalogue*.[7] Mere numbers go some way to establishing the pervasiveness of the myth, so it is worth noting that, up to 1982, Glut lists 130 other fictions based on *Frankenstein*, almost 50 fiction series, more than 40 film adaptations, over 80 stage productions, and 80 films. In the comics, the only area where Glut claims to have been selective because of the overwhelming mass of material, he lists nearly 600 individual items and 30 series, not counting newspaper strips.

As with popular culture in general, there is far too much material here to get to grips with. Widening the scope just a little inflates Glut's numbers still more. Stephen Jones' *Illustrated Frankenstein Movie Guide*, compiled ten years after Glut's catalogue, lists more than 400 films more or less based on the *Frankenstein* script.[8] But it is worth going beyond the first plausible generalisation – that *Frankenstein* has been continually retold in all possible media since its first appearance – to outline the main points in the wider history of the tale since 1818. After that, we must return to the question of what gives this particular story its appeal.

Beyond the book

When Mary Shelley published her anonymous novel in 1818, it was greeted as something of a scandal by reviewers, though Sir Walter Scott in the *Edinburgh Magazine* famously discerned some merits in the tale.[9] The fact that the story was described as blasphemous, offensive to morals or repulsive did not necessarily deter the middle-class novel-reading public. But even with the increasing popularity of circulating libraries, the potential readers for an expensive production like the first edition, in three volumes, probably numbered only a few thousand.[10] The reading public was in the midst of a phase of rapid expansion, of which the Gothic boom was an integral part, and later editions would attract many more readers.[11] Nevertheless, to truly begin its other life, the story needed to break the bounds of the printed page, and this it duly did with the first stage production. Richard Brinsley Peake's *Presumption; or the Fate of Frankenstein*, first performed in 1823 in London, was the first in a long line of nineteenth-century dramas based on the book.

Any tendency today to ascribe the widespread familiarity with *Frankenstein* solely to the power of the twentieth-century cinema is quickly dispelled by Steven Forry's study of the propagation of the tale.[12] Peake's play caused a sensation, prompting both the first reprint of the novel and a succession of further dramatisations – fourteen in England and France in the next three years alone. Forry helpfully suggests that there were three movements in the dissemination of the myth. From 1823 to 1832, it was transformed for popular consumption. Between 1832 and 1900 the story diffused more widely in its simplified form. And from 1900 to 1930 there was a transition, as drama was gradually overtaken by cinema: the indelibly memorable James Whale film of 1931 was in fact based on a recent stage play, not on the novel.

The transformation took effect rapidly. Within three years of the first performance of Peake's melodrama, the main lines which stage plays would follow were set. The demands of the stage led dramatists to concentrate on the Gothic background, demonising the creature and expanding the laboratory, giving it stronger alchemical overtones. They reduced the complex, troubled creator to a simpler Promethean figure, challenging the gods and getting the expected comeuppance.

The early stage productions also served to introduce 'Frankensteinian' as an adjective, a usage recorded by Forry in *The Times* within a month of Peake's première. They also began a process in which the story, in much-reduced form, became familiar to much wider audiences. Some saw plays, which continued to proliferate. Some read new versions in print.[13] Others read the original tale in one of its numerous reprinted editions, generally of the 1831 revision. And the image of a monster was linked to Frankenstein by political cartoonists through the rest of the century, as Forry documents.

Mary Shelley commented obliquely on this process, one suspects, when she bid her 'hideous progeny' go forth once more in the preface to the 1831 edition. She had seen Peake's play, and rather approved of its effect on the audience. Her phrase playfully acknowledges her well-developed awareness of the analogies between literary and biological parentage, her monstrous story about a monster. Frankenstein's creature never had his chance to spawn a new race of beings, but the story has been fertile beyond all expectation.

We shall return to other nineteenth-century literary productions which drew inspiration from *Frankenstein,* but for now I will sketch the later cinematic history which so dominates our ideas of the story in this century. It is a critical commonplace to say that cinema is the pre-eminent vehicle for the propagation of myth in contemporary mass societies. This is true both of the way film narratives are told, and of the choice of stories.[14] It is certainly true in the case of the *Frankenstein* myth.[15] The Edison Company made a silent movie of *Frankenstein* as early as 1910, of which only stills now survive outside private hands, but the three most important productions of the story are the British director James Whale's film for Universal in 1931, Terence Fisher's *Curse of Frankenstein* for Hammer in 1957, and Kenneth Branagh's *Mary Shelley's Frankenstein* in 1994. Here I will briefly consider the first two, deferring comment on Branagh until a later chapter.

Both are recognised as key contributions to the horror genre, although

Whale's film, in particular, also appears prominently in histories of science fiction film. And both have received a good deal of critical attention, although not yet on the same scale as the novel. Such attention establishes that these are complex cultural products in their own right, not merely stripped-down versions of the novel. The details are fascinating, though they can only be sketched here. More important is the general power of cinema, as a mass medium with its own production demands, to act as a cultural amplifier. This effect begins with Whale's film itself, but goes far beyond it.

The critic Paul O'Flinn suggests that there are three types of shift to consider between book and films: shifts of medium, audience and content. First, the medium dictates the loss of the embedded narratives of the book, which he claims cannot be filmed. The middle-class audience for a Gothic novel is replaced by a mass audience, and 'the radical change in the class nature of producer and audience hacks away at the content of the original, so that the book is reduced to no more than an approximate skeleton, fleshed out in entirely and deliberately new ways'.[16] But all these processes had begun more than a century before, in drama and cheap fiction. So what was actually new?

Big business is part of the answer. The 1931 *Frankenstein* was made by a Hollywood studio not then in the first division, which had just had a great success with *Dracula*, starring Bela Lugosi. Looking for another well-known story, the studio decided that the idea which had horrified Mary Shelley would still gratify audiences, and bought the rights to the play recently staged successfully in London, *Frankenstein: An Adventure in the Macabre*. The film was made in two months and earned the studio $12 million for a $250,000 outlay, confirming the public's appetite for horror.

As to the story, it opens, not with Walton but with Henry (not Victor) Frankenstein and his assistant Fritz disinterring a recently buried corpse from a graveyard. They are already embarked on their project, gathering material.[17] The creation scene comes after we have seen Fritz steal a 'criminal brain' from the medical school, and takes place in the presence of the other main characters – Frankenstein's fiancée Elizabeth, their friend Victor and Henry's former mentor Dr Waldman. Then comes success, and the balance of the film shifts as Boris Karloff's monster dominates the screen. First abused by Fritz, then drugged by Waldman, he escapes. As in the novel, he returns to Frankenstein on his wedding day, then flees again, pursued by a village mob, to meet a fiery end in a blazing windmill.

There is much more to be said about the construction of the film,[18] but what were its main effects on the imagery associated with the *Frankenstein* myth? The simple moral of the earlier plays is heavily underlined, as emphasised by the prologue addressed directly to the audience which warns them that they may find the tale shocking, because 'We are about to unfold the story of Frankenstein, a man of science, who sought to create a man after his own image, without reckoning upon God. It is one of the strangest tales ever told. It deals with the two great mysteries of creation: life and death.'

Aside from this, what are the most memorable images? One is the creation scene, at the height of a storm in Frankenstein's magical, electrical, but still scientific laboratory. Before Victor, Elizabeth and Waldman, he exposes the body he has created to 'the great ray that first brought life into the world'. At the end, as the creature's hand lifts from its side, he exults repeatedly: 'It's alive!'[19] The other enduring image, needless to say, is Karloff's mute, pathetic, but menacing creature, still his creator's *alter ego* but balanced against a rather weaker character in this film than in the novel. This is the image of Frankenstein in a thousand comic books and cartoons, and the one which confirms the identification of the name with the creation, not the creator.

One striking film can give a story a reach possible in no other contemporary medium, but further amplification of the myth was only begun by Whale's first *Frankenstein* feature. Universal built up the horror genre by creating cycles of films. This was a deliberate business strategy, both giving audiences more of a tested product, and saving money by recycling sets, costumes, and even film footage.[20] Just as Mary Shelley's doctor learned how to reanimate a patchwork corpse, so the cinema would have to resuscitate the monster after its apparent death at the end of Whale's first film.

The highly successful sequel to the 1931 film was *Bride of Frankenstein* (1935), which claimed to turn to the novel for inspiration, even featuring Elsa Lanchester (who played the female creature) as Mary Shelley in a prologue. It adds another scientist, Dr Pretorius, who has succeeded in creating life independently, and who coerces Frankenstein into joining forces with him so that they may perfect their work together. There followed *Son of Frankenstein* (1939), and *The Ghost of Frankenstein* (1942). Still later films such as *House of Frankenstein* and *Frankenstein Meets the Wolf Man* (both 1946) introduced other stock Universal monsters into their increasingly crude stories – Dracula in the first, the Wolf Man in the second.[21]

As this suggests, the importance of the 1931 *Frankenstein* went far beyond even its growth into a series. Along with *Dracula* it established one of the archetypes for a whole film genre, the horror movie. This is not simply a matter of a multitude of further Frankenstein films down the years. It is more that the basic narrative of the *Frankenstein* myth is at the heart of many other films as well.

Here, plot summaries become repetitious, and we need numbers. These are helpfully supplied by Andrew Tudor's detailed analysis of 990 horror films distributed in Britain between 1931 and 1984. He suggests, most generally, that such movies follow a three-part narrative pattern: instability is introduced into an apparently stable situation; the threat to instability is resisted; the threat is removed and stability is restored. Within this general scheme, he identifies three sub-types, one of which is the 'knowledge narrative', more familiarly known as the 'mad scientist' story. In such a narrative, someone, most often a scientist, uncovers knowledge which poses a threat to the everyday ordered world, either deliberately or by some accident in its use. There is then an interval of rampage or destruction, with heroic efforts to resist or destroy whatever is doing the damage. In the closed version of the narrative, which is by far the most common, order is eventually restored and the boundaries between known and unknown safely secured again.[22]

Frankenstein is plainly the most familiar expression of this story, but its prominence in the wider genre is extraordinary. Tudor finds that between 1931 and 1960, a period covering some 260 films, between a third and a half of all the stories were of this type. After this time the proportion declines, although the numbers are still appreciable as the genre grows substantially. In particular, he emphasises how characteristic mad science (though often with partially redeeming good intentions) is of the films of the 1930s and 1940s. The preoccupations of these movie scientists were almost always biological – creating man-like creatures, restoring life to the dead, creating living tissue, or manipulating evolution to create ape-humans.[23]

Describing films like this in any detail is redundant. There is very little thematic variation.[24] But it is important to note their proliferation as part of any full account of the *Frankenstein* myth. More central, though, is the next major cinematic realisation of the story, Hammer Films' *The Curse of Frankenstein* in 1957. As in 1931, this came from a relatively small studio looking for a lucrative way to exploit the changed state of the art in film production. Hammer Films had had considerable success in 1955 with the horrific science fiction story, *The Quatermass Experiment*, a black and white remake of a BBC Television series. The company decided to build on this by returning to established horror stories, and going into colour production. *The Curse of Frankenstein*, made in 1956 and released the following year, was their first choice.

From this distance the film was most notable for focusing attention on the creator (now Baron Frankenstein), rather than the creature. This was partly an expedient reached because Universal had rights to Karloff's depiction. The script made the Baron more ruthless than previous Frankensteins. He carries on an affair with a servant girl while wooing Elizabeth, and does not dispatch an assistant to steal a criminal brain but personally procures a high-quality product by murdering his elderly colleague Professor Bernstein. The actor who became famous for playing him, Peter Cushing, saw him as a misunderstood pioneer, claiming that he always based his playing of Frankenstein on Robert Knox – the Edinburgh surgeon who was vilified for buying corpses from Burke and Hare.[25]

As well as including the compelling figure of Cushing's aristocratic, sarcastic Frankenstein, the first Hammer film also went further than previous films in special effects, and the colourful spectacle which resulted was too much for some critics. As with the 1931 film, the extreme imagery we are now inured to, and the high technical brilliance and enormous resources of later cinema, make it hard to recapture the impact of *The Curse of Frankenstein* on contemporary audiences. Some of the negative reviews give the flavour, one in keeping with common reactions to Mary Shelley's novel in 1818. The *Daily Telegraph*, for example, suggested that 'when the screen gives us severed heads and hands, eyeballs dropped in a wine glass and magnified, and brains dished up on a plate like spaghetti, I can only suggest a new certificate – "SO" perhaps; for Sadists Only'. Others described the film as repulsive, revolting, depressing and degrading.[26] They responded, in fact, in the same fashion as the *Quarterly Review* had responded to the novel, when its reviewer in 1818 found that

it inculcates no lesson of conduct, manners or morality; it cannot mend, and will not even amuse its readers, unless their taste have been deplorably vitiated – it fatigues the feelings without interesting the understanding; it gratuitously harasses the heart, and wantonly adds to the store, already too great, of painful sensations.[27]

This denunciation is generally ascribed to anti-Shelleyan political feeling, but the similarities with responses to Fisher's film perhaps suggest other sources for the repugnance the tale evoked. A further similarity is that, as the cultural historian Peter Hutchings points out, by 1957, as probably in 1818, this kind of condemnation was an effective advertisement for a film which was not, after all, aimed at readers of the *Daily Telegraph*. Other reviewers recognised that there was an audience for films of this type, and were either indifferent to Hammer's gore or amused by it. All of them helped it on the way to major commercial success, in Britain and the USA, and cemented Hammer's commitment to the genre. The company went on to revive *Dracula* and *The Mummy* as well, and also emulated Universal by constructing horror cycles. Their first *Frankenstein* was followed by six more, all but one of them starring Peter Cushing, the last in 1973.[28]

Hutchings' account makes it clear that, as one would expect, the second main *Frankenstein* cycle was shaped by its own context. A full analysis of the films would take in the state of the British industry, the changing audience, the shifting climate of censorship, and those questions of identity, authority, gender and class which were troubling British culture of the 1950s and after.[29]

But while these features of the context of production and reception are important, it is equally intriguing that it was again *Frankenstein* which played a central role in the evolution of the horror genre. And, as with the Universal cycle and its reflections and imitations, this also maintained the *availability* of the simple *Frankenstein* script. Its continued identification with the newly-shocking, or the barely acceptable, ensured that the disquieting qualities of that script were perpetuated as well.

There are many more instances of cultural products constructed around the *Frankenstein* script than I can realistically introduce here, as Glut's catalogue reveals. But this brief survey shows two things above all: the adaptability of the basic story, and its ability to retain its essential elements even as it inspires more retellings. It is precisely this which is best summed up by saying that the story has been transformed into myth. Having said that, it is time to look a little more closely at what has been retained throughout that process. Then we will ask again what the continuing appeal of the myth might be.

Science, modernity and the body

The accumulated retellings of the *Frankenstein* myth are now so numerous as almost to defy empirical analysis. Today, we encounter Frankenstein in many forms. Any of the old films may still be seen as late-night TV fillers, or on video. There are even two films which incorporate versions of the origin myth of the novel, mixing together the story of Mary, Percy and Byron by the lakeside with the creation of the monster.[30] New films continue to incorporate elements of the story, from *Demon Seed*, in which the monster is a computer which finds a way of inseminating a human female, to *Robocop*.[31] Numerous editions of the novel remain in print, and new variations on the story continue to appear in printed fiction. Some of these, like Steven Gallagher's *Chimera*, are filmed in turn. Others, like Hilary Bailey's striking *Frankenstein's Bride*, remain as solely literary efforts. I will return to some of these most recent renditions in the final chapter, but it is clear that the form of the story still appeals to artists in several media.

In addition, as with all truly frightening myths, we have tried to tame *Frankenstein* by making fun of it. Karloff's monster has been domesticated, in media ranging from the 1960s US television series *The Munsters* to the British children's comic the *Beano*, which features Frankie Stein.[32] A distant descendant of Karloff even featured as Frank in the British Conservative government's television commercials for shares in its soon to be privatised electricity generation concern in the early 1990s. This taken-for-grantedness shows how well the cultural script has been learned. In consequence, the single word 'Frankenstein' is seen constantly as a metaphor in media commentary of all kinds, especially political commentary.[33]

Why, then, has the story endured? Is it simply because the frame is so open at various points that it is infinitely adaptable? Or are there particular reasons, culturally general enough to read across all the retellings, with all their differences of detail, yet still specific to the culture which we share with Mary Shelley – broadly, the culture of modernity?

The first answer is to try to isolate what has endured in all the renderings of the myth since 1818. The story, for all its familiarity, is still a frightening one. It is frightening because it depicts a human enterprise which is out of control, and which turns on its creator. So much carries over from the earlier myths about the getting of knowledge. But *Frankenstein* is about science. What is more, the science is pursued, if not always with the best of intentions, then for motives with which we can readily identify. In the most striking retellings, the myth is never a straightforward anti-science story. There is something admirable about Victor Frankenstein, about Henry Frankenstein in James Whale's film, even about Peter Cushing's Baron Frankenstein. Even so, our sympathies are always torn between Frankenstein and his monster. The *Frankenstein* script, in its most salient forms, incorporates an ambivalence about science, method and motive, which is never resolved.

The retention of science in all the later derivatives of the story is the most striking feature of the myth. After all, in the original text, once Victor's narrative begins, the creation of the monster is accomplished in a scant thirty pages, in which space is also given to the background and education of the monster's creator. The scientific details are few. After Victor's 'brilliant light' dawns we never learn more than how he eventually 'collected the instruments of life around me, that I might infuse a spark of life into the lifeless thing that lay at my feet'. Yet it is those first thirty pages that supply the seeds of almost all of the images derived from *Frankenstein* which appear in so many variations in later stories about science and scientists.

Among others, we can distinguish in *Frankenstein* models for the scientist whose good intentions blind him to the true nature of his enterprise: 'wealth was an inferior object; but what glory would attend the discovery if I could banish disease from the human frame and render man invulnerable to any but a violent death!' Victor proclaims. And so say all of us mortal readers. But Victor also personifies the scientist as Faustian knowledge-seeker; 'the world was to me a secret which I desired to divine', he remembers, and he recalls that 'none but those who have experienced them can conceive of the enticements of science' or, as a narrow materialist, 'On my education my father had taken the greatest precautions that my mind should be impressed with no supernatural horrors . . . a churchyard was to me merely the receptacle of bodies deprived of life.' There are also hints that science has some drive of its own, external to the will of the scientist and eventually overwhelming him. 'Natural philosophy,' Victor reflects sadly, 'is the genius that has regulated my fate.' Amidst all the simplifications, deletions and elaborations of the original, the identification of Victor as a scientist has remained inviolate. It is science which gives him his success, and that success gives him power over life. Even though his character was first drawn before biology was a separate discipline, Frankenstein is always a proto-biologist.

So the endurance of the myth plainly does testify to a deep disquiet at the potentialities inherent in scientific discovery in general, and the science of life in particular. And it is a disquiet which Mary Shelley appears to have tapped into at a remarkably early stage in the development of modern life science. The appearance of the story, and its ready acceptance, so soon after Erasmus Darwin's speculations were published, suggest that unease at the prospect of science attaining powers over life is readily evoked in the public mind. So I agree with all those who have suggested that the *Frankenstein* myth both expresses and reinforces an undercurrent of feelings about science; that in George Levine's phrase, it 'articulates a deeply felt cultural neurosis'. But what, exactly, does this neurosis consist of?

It is clear that what we now call biomedical science, or the possibility of a technologically effective biology, has played a key role in shaping the modern attitude to science. We have always been prisoners of the body, victims of morbidity and mortality, and we desire the power that biology might give us to relieve these burdens. In more recent times, this can be seen from other kinds of evidence. Medical and biological stories have long accounted for a large proportion of the press reporting of science, for example.[34] Editors appear to regard such stories as of more interest to their readers than other scientific items. The news-consuming public, in consequence, may be more aware of events in biological science than in other fields.[35] The nature of their interest has also been long established. Turn the pages of a major newspaper from the early years of this century, the *San Francisco Chronicle*, say, and you will find front-page stories on radical new surgical procedures, on the possibility of choosing the sex of a baby, on proposed scientific techniques for prolonging life, and on putative cures for cancer. These stories show the early convergence of news values and the territory of biomedicine. Biological research and medical practice mean birth, sex and death; suffering, disease and disability.

Biologists who become visible to the public are aware of the hopes and fears their science raises. As the French geneticist and popular writer Jean Rostand – of whom we shall hear more – attested in the 1950s: 'The best way to gain an idea of what the human, *emotional* value of biology can be, is to look through some of the strange correspondence that a biologist receives . . . people take him for a magician, a healer, a confessor, a friend.'[36]

Among the letters he describes are those from couples seeking to replace a lost child with a perfect twin, queries on the consequences of mixed marriage, people seeking confirmation of paternity by blood typing, enhancement of their children's intelligence, rejuvenation for the elderly, sex changes or cures for infertility. Rostand concludes that 'the science that provokes such appeals, prayers and confessions, the science that penetrates into private life, and whose warnings or advice can influence a marriage, a decision to have children, a person's destiny, is no ordinary science'.[37]

These examples express very well the idea that biology is indeed 'no ordinary science' for the public. It is the science which touches on the most potent wishes of human life. The realisation that biology offers the prospect of ultimate control over or transformation of the living realm, just as physical science controls and transforms the physical environment, thus evokes deep-rooted feelings. This realisation by itself can produce either positive or negative reactions, and *a priori* both may seem equally likely. Indeed, as I shall discuss in the next chapter, the first great practical successes of scientific medicine produced more general endorsement of the virtues of science than ever before. Yet, the response to a genuinely effective experimental biology has always had its negative side as well. To understand *Frankenstein*, we need to look further into the roots of that ambivalence.

One set of clues here comes from early responses to industrialisation, and to the conspicuously powerful applications of physico-chemical technology which it highlighted. These responses were bound up in complex ways with changing awareness of the natural world, and particularly of the body. A number of writers have suggested that the body is the locus for a set of feelings which play a major part in determining responses to the effects of industrialisation.

The transforming potential of physico-chemical technology was made manifest in the most concrete way in the eighteenth and nineteenth centuries, and the evidence was beginning to be apparent to all when *Frankenstein* was written. The impact of technology differentiated the industrial revolution from the past in ways that were hard not to notice, as the machine transformed entire landscapes.[38] The ensuing cultural changes, which expressed or embodied responses to the vastly expanded role of technology while these changes were actually taking place and first being recognised as being revolutionary in scope, are complex and hard to summarise. But works like Francis Klingender's pioneering *Art and the Industrial Revolution*, with its record of the iconography of industrialisation, or the critic Leo Marx's study of industrialisation and the pastoral ideal in America, *The Machine in the Garden*, map out some sectors of creative activity in which these changes were reflected, discussed, evaluated and interpreted.[39] A later study by the anthropologist Jonathan Benthall, partly inspired by them, offers a typology of responses to technology.[40] His classification ascribes particular importance to the natural world, and especially the human relation with that world represented by our biological make-up.

There is a contradiction between our spontaneous, visceral experience of the world, mediated through the senses, and our technological capacity to order the world through analysis and measurement. Faced with this contradiction, one possible response to the objectification of the world by science and technology is a 'recoil to the body', a reaffirmation of the intensity of bodily feelings and of the integrity of the person. Thus the body, as an image of the natural world in microcosm, becomes a metaphoric resource for those who criticise the scientific world-view.

Studies of the responses to the machine – as a symbol of modern technology – in Victorian literature, find that there is a consistent pattern of opposition between the mechanistic and the organic.[41] Mechanism and machines are employed to represent the scientific mind, while the organic represents a set of qualities associated with the living organism. Mechanism implies fixed causality, determinism, lawfulness, all of which are opposed to the irregularity, wholeness and intuitive harmony supposedly found in the natural world. In all of these studies, images of the natural world are an inspiration for opposition to the scientific world-view and to the effects of the objects and artefacts deriving from that science.

But a science of life put this strategy in question. At least since Descartes, the most conspicuous proponents of a scientific study of living organisms have abhorred vitalism and espoused mechanism. If those who viewed physical science with distaste sought to retreat to the comforting haven of the living world, science sought to mechanise even that. Here is one reason for the bitterness of the recurrent disputes between mechanists and vitalists, and also for the belief that perceptions of the likely development of biology may have played an especially large part in sustaining ambivalence about science. The growing abstraction of biological work can become a symbol, for some, of the estrangement of man from nature brought about by industrialisation.

There is a yet wider dimension to this history, and to biology's potential significance. Taking its measure involves a broader account of the changes which were in train during Mary Shelley's early life, changes which are not completely captured by speaking of the industrial revolution. They are the changes, of which startling technological progress was only one element, that marked the beginning of modernity. We can see how the creative imagination could intuit the ultimate implications of these wider changes by looking at a reading of another myth, one closely related to *Frankenstein*. The reading I want to examine is Marshall Berman's study of *Faust* as a myth of modernity.

In the introduction, I summarised Berman's view of the experience of modernity as one of forces unleashed, of unlimited change in all spheres of life. He proposes Marx as one of the first to recognise the full extent of these changes, because the bourgeoisie, while proud to be the agents of change, were functionally blind to the fact that it would eventually sweep away all that they had created. Berman's title, with its half-echo of Prospero, is taken from one of the more striking passages of the *Communist Manifesto*, in which Marx offers a hymn to the unstoppable power of capitalist competition:

> Constant revolutionizing of production, uninterrupted disturbance of all social relations, everlasting uncertainty and agitation, distinguish the bourgeois epoch from all earlier times. All fixed, fast-frozen relationships, with their train of venerable ideas and opinions, are swept away, all new-formed ones become obselete before they can ossify. All that is solid melts into air, all that is holy is profaned, and men at last are forced to face with sober senses the real conditions of their lives and their relations with their fellow men.[42]

Before Marx, Berman tells us, the most penetrating analysis of the course of modernity came from Goethe, and can be found in his extended version of the Faust myth. Goethe's *Faust*, begun in 1770 and finished sixty years later, when Goethe was 82, is the work of a writer of far greater powers than Mary Shelley, and is justly celebrated as one of the great works of European literature. For Berman, the work, and especially *Faust Part Two*, is a reflection on Goethe's own experience of the second phase of modernity: 'the work was in process all through one of the most turbulent and revolutionary eras in the history of the world . . . the whole movement of the work enacts the larger movement of Western society'.[43]

In Part Two, which Berman calls the tragedy of development, Faust uses the power granted him by Mephistopheles to transform a whole landscape. He is no longer satisfied with the good things sought by earlier Fausts. Money, sex, power and glory are not enough. He wants to refashion the world, and creates the technological and social organisation to do it. The whole of Part Two, according to Berman, is a parable about modern progress, and 'Goethe's point is that the deepest horrors of Faustian development spring from its most honorable aims and its most authentic achievements'.

Now there is much to be said for Berman's reading of *Faust*, and it is certainly true that the story is still one of our most powerful myths. But, when we compare it to *Frankenstein*, it is evident that Goethe's modern myth is only a partial transformation of the older stories of Faust. Mary Shelley, writing in the same period, and confronting many of the same concerns, created a story which is more deserving of the title of a modern myth.[44] While Faust, for all his recognisably modern outlook, still relies on a pact with the devil to secure his power, and pays the price, Frankenstein achieves his powers unaided. The Englishwoman's story is a modern myth because it is grounded in science, not the supernatural.[45]

That, though, is not the only significant respect in which she differs from Goethe, in spite of all they share in terms of intellectual influence and sensibility. Faust is preoccupied with operating on the world, with physical and social transformation. Man will be transformed in the new world, but only through all the other changes Faust brings about. Frankenstein, on the other hand, is set on transforming humans directly. If he can discover the secret of life, then he can father a new species. To do so, he will experiment directly on the body.

Here, I think, Mary intuited the power of a threat which would come to seem graver as time went by. In a world where everything appeared to be subject to change, where it was becoming apparent that 'all that is solid melts into air', there was one sphere of existence which was exempt. The natural world, although it could be reshaped by physical onslaught on the landscape, although it could be despoiled or laid waste, was not yet open to technological manipulation. The forms and varieties of creatures, the hierarchy of species, the biological imperatives of existence, were fixed points in an ever-changing world.[46]

The human body, too, as I have suggested, provided an unchanging ground for experience of other changes. This does not mean, of course, that *experience* of the body, or ideas about its constitution, did not change.[47] But the body itself was not seen as changing by those experiencing the first rush of modernity. While the dead body had been anatomised for two centuries, in pursuit of a science of the interior which had made a deep impression on Renaissance and early modern culture, that science was still largely descriptive. The living body was not yet susceptible to the kind of science being developed in other areas, in which 'the object to be known . . . will be known in such a way that it can be changed'.[48] Frankenstein the character, and *Frankenstein* the novel, are both steeped in the anatomical tradition. But this anatomist goes further. Mary Shelley made the necessary imaginative leap, and fashioned an image of a science working on the body to transform it, a science which might one day come to pass. Now that we are indeed building such a science, we can see that it has always been a part of the modern project. She saw this right at the start. If, as Berman says, Goethe's key insight is the ambivalence stemming from the fact that 'the deepest horrors of Faustian development spring from its most honorable aims and its most authentic achievements', then the best horror story would be rooted in the power which we simultaneously most desire, and most dread: power over the body. *Frankenstein* focused attention on that prospect nearly two centuries ago. We still feel the pull of the story because that power is now ours for the asking.

Chapter Three

As Remorseless as Nature: the Rise of Experimental Biology

The number of persons . . . who are becoming biologists without becoming medical men, is very much increasing. Modern civilization seems set on acquiring, almost universally, what is called biological knowledge; and one of the consequences of that is, that whereas medical men are constantly engaged in the study of anatomy and physiology for a humane purpose . . . there are a number of persons now who are engaged in the pursuit of these subjects for the purpose of acquiring abstract knowledge. This is quite a different thing. I am not at all sure that the mere acquisition of knowledge is not a thing having some dangerous or mischievous tendencies in it . . .

> Sir Henry Acland, testimony to the Royal Commission on the
> Practice of Subjecting Live Animals to Experiments for
> Scientific Purposes, 1876
> Quoted in French, 1975

In 1933, the London publisher Odham's produced a lavishly illustrated popular account of biology, *The Miracle of Life*. For the frontispiece they chose, not one of the wonders of nature, but a portrait of the modern researcher. Anonymous, a middle-aged man in an immaculate white coat sits in front of a binocular microscope, intent on the brightly lit stage; in his hands, a pair of dissecting needles; around him, reagents and glassware.

He is, in fact, the stereotypical biologist of the first half of the twentieth century. His quest, the caption declares, is nothing less than the secret of life. Somehow, his furrowed brow and concentration on manipulating we-know-not-what are unsettling. What is the 'matter' he is 'scanning'? Where did it come from? Is it still alive? What will become of it afterwards? All these questions, as we shall see, were as readily evoked then as now.

This biologist's portrait is the midway stage in the iconography of investigation in the life sciences over the last hundred years, which develops roughly in three phases. In the mid-nineteenth century, a Haeckel or Huxley would sit for a portrait in ordinary dress, perhaps in a book-lined study or by the lectern, and usually hefting a primate skull. The biologist, poised between his past incarnation as a gentleman amateur and his new-found professional status, had the measure of man, or ape. But the emphasis of the image was more on morphology and taxonomy than experiment. The skull was all but inanimate and could be described and classified by the unaided eye, but not operated on or manipulated.

By the 1890s, the scene changes. The scientist, like Odham's biologist, has donned a white coat and moved into the laboratory. He is armed with a microscope and glass dishes or test-tubes.[1] This second image persisted with remarkably few changes through the first half of this century, until the elucidation of the structure of DNA. After the mid-1950s, the characteristic portrait changed again. The white-coated figure is now most often seen posed with a molecular model or a blackboard diagram. The physical scientist had his equations or his computer, the biologist the double helix. More recently, as a development of the same phase rather than a new departure, the object of study is depicted as a stretch of printed DNA code, or a photostrip of a DNA fingerprint – the emphasis on information rather than structure.[2]

Although the images have changed, some of the messages they convey have remained constant. All these biologists work in realms distant from the public of their time, but the distance has increased – and not only sartorially. The portraits symbolise two shifts in the practice of biological scientists, from description and classification to experimentation, and from study of the organism to the abstraction of biochemistry, genetics and, ultimately, molecular biology. 'Life' has become less recognisable as the work done in the private space of the laboratory has become less comprehensible. With each shift, the matter under investigation has become further and further removed from everyday experience, and from the whole organism. Today, the essence of each creature, including human beings, is represented as a matter of encoded information – the body is merely a vessel for the propagation of DNA.[3]

At the same time, growing abstraction goes hand in hand with increasing control. The information, and hence the details of the organism can be read, edited and rewritten, according to the current governing metaphor.[4] The aim and object of biology is now knowledge which enables manipulation. The images show that deeper knowledge of living systems is being sought, and that in some senses the search is successful. We really can transfer genes between organisms, modify the code, or command cells to manufacture proteins to order. The meaning of that success is something we are still debating.

The opening chapters of this book examined the origins and subsequent history of the *Frankenstein* story, a modern myth about biology. We saw how Mary Shelley's novel was grounded in the science of her day, but my sketch of the later uses of the *Frankenstein* script was drawn without any attention to later scientific developments. Having dwelt on the myth, the remainder of the book pays more attention to science, but still concentrating on the ways in which science and scientists were represented in public. I will offer a very broad-brush account of the investigation of life since the turn of the nineteenth century, a history of increasing control over the living world conceived as a set of objects for experimental study. Interwoven with this story is another, equally significant tale, of the development of public images of life science. Here, we will meet Frankenstein and his descendants again, as well as examining other stories which have made their mark on the social representation of biology. And we will see how specific changes in science have affected (and been affected by) these representations. In the next chapter, the focus is on a particular episode at the start of this century, when it appeared for a while that the creation of life in the laboratory was about to turn from fiction into fact. The task of this chapter, however, is to consider some key trends and developments of the nineteenth century, and view their effects on the response to biological science. In particular, it is concerned with the transition between the first image of the biologist, the frock-coated skull-hefter, and the second, the white-coated microscopist; the transition from natural historian to experimenter.

When Mary Shelley wrote *Frankenstein*, science was still the province of the gentlemanly amateur. Biology was not yet a separate discipline, still wedded on the one hand to medicine and on the other to the more general pursuit of natural philosophy. Although she used a biological image to animate her myth of modernity, the science of her time was centred in chemistry and physics. And Humphry Davy, with his laboratory in the basement of the Royal Institution, was the exception among scientists, most of whom still needed private means to support their research.

We can remind ourselves where science stood just after *Frankenstein* by looking briefly at a claim for laboratory-created life which arose a few years after the publication of the third edition of the novel in 1831. Andrew Crosse, a young West Country gentleman of means, a radical and a devoted experimenter, became briefly famous in the wake of press reports of what he had achieved working alone in his somewhat disorganised laboratory. Crosse, who was mainly interested in the influence of electricity on crystal formation, was passing currents through mineral solutions for long periods. On seeing small mites emerge within the apparatus, he believed that he had witnessed the electrical creation of life. This extraordinary episode has been examined in detail by the historian James Secord.[5] He notes that the advent of the steam press had transformed the reach of newspapers and magazines and there was now, for the first time, a media system delivering news rapidly to a nationwide public.[6] From now on the wider public heard about science from sources distinct from those used by the practitioners.

In consequence, scientists began trying harder to influence or control the public image of science in the media outside their own hands – attempts which have gone on ever since. The British Association for the Advancement of Science, founded in the same year as the third edition of *Frankenstein* appeared, played a key role in the controversy over Crosse's experiments. In 1836, his work on mineralisation was fêted at the BA's annual meeting, so that he was seen as 'a genius whose authority had been recently validated by the scientific elite'. But the metropolitan scientists were nonplussed a few months later when local newspaper reports of Crosse's mites were taken up by *The Times*, and then by virtually every other newspaper and magazine in the country.[7]

For most scientific observers, the electrical experiments were reputable, but the generation of life was not. The boundaries between life and matter, and between natural law and divine experiment, were closely guarded. Some denounced Crosse as an atheist, and called for the destruction of his laboratory. For them, Secord suggests, 'Crosse was nothing less than a Frankenstein, a lost soul wandering beyond the proper limits of human knowledge'. Others welcomed his results as the foundation for a new understanding of origins, based on recreating the conditions under which God had ordained that creation could occur.

As fellow electrical experimenters, geologists, taxonomists and zoologists argued over the significance of Crosse's results, and what, if anything, he had produced – cheese mites appeared to be the answer – most attempts to replicate his experiment failed. It was these failed attempts which were announced with a flourish at the next year's British Association meeting. But there were enough reports, founded and unfounded, of experiments which backed Crosse to keep the story of the mites alive for many years.[8] In the end, though, whatever impression they left with the public, Crosse's claims had to contend with the fact that, as Secord puts it, 'studies of electricity and life had no place on the disciplinary map of Victorian science'. Claims about the possibility of creating life under scientifically controlled conditions which would be taken seriously by scientists *and* the public would have to wait for a new century, and a different kind of science.

Secord describes the Crosse controversy as a struggle over the significance of experiment. But in the life sciences at this time, experiment was still the exception. As science continued its transformation through the remainder of the nineteenth century, the professionalisation and differentiation of disciplines which was already under way in the 1830s proceeded apace. By mid-century, biology was recognised as a discipline in its own right. And if it was to take its proper place among the sciences, it must be an experimental discipline. Alongside natural history, taxonomy and comparative anatomy, with their traditions of observation and classification, there grew a commitment to analysis of biological mechanism through experiment. Amid all the other changes of these years, the years in which modern science took on many of the features which we can still recognise, it was the experimental practice of biology which was among the most powerful influences on the public image of science. By the end of the century, a reading public also transformed since Mary Shelley's day was enjoying Stevenson and Wells, who we can now see created the next two important additions to the stock of fictional images of scientists with a lasting presence in our culture, Dr Jekyll and Dr Moreau. But if science had been altered out of all recognition, perhaps the response to these changes mainly served to deepen the grip of the existing image, to dress Frankenstein in new clothes.

In biology, the rise of experimentation began in France and Germany in the 1820s and 1830s. The professionalisation of continental chemistry helped inspire similar changes in animal physiology, with the creation of prestigious university posts and the formation of research schools.[9] The pioneers were Johannes Müller in Berlin and François Magendie in Paris. Müller trained his pupils in use of the microscope and physico-chemical techniques, while

Magendie emphasised the use of chemical methods – and vivisection. As their students in turn began research and teaching, they joined a wider effort to make biology an experimental science.

They promoted that effort in at least two ways. First, the experimentalists began to write manuals of procedure, most famously Bernard's *Introduction to the Study of Experimental Medicine*, first published in French in 1865.[10] Secondly, they were often violent antivitalists. The new breed of physiologists asserted that life could be explained in purely physico-chemical terms.[11] Both the manuals and the metaphysics were to evoke responses from non-scientific publics. Bernard, above all, was an unapologetic vivisectionist, and it was objections to the *methods* being used in the new physiological laboratories which became a focus for public discussion. Vivisection was not new, as Victor Frankenstein testified when he told Walton how he 'tortured the living animal to animate the lifeless clay'. But its large-scale use was.[12] It was now an integral part of an interventionist, empiricist approach to biological problems. After mid-century, ether anaesthesia encouraged those with a less Cartesian view of the animal machine to overcome their scruples and adopt the new techniques.

The popular reaction was most vocal in Britain, which at once cautions against any assumption of a simple reaction to cruelty to animals behind the closed doors of the laboratory. Although the first experimental physiologists were Germans and Frenchmen, and the number of animal experiments performed in their countries remained higher than elsewhere until near the century's end, the movement against vivisection found more fertile soil across the English Channel.[13]

Britain lagged behind the continent in experimental physiology for several reasons, including a relative neglect of science, the fact that universities were less open to innovation, and a conservative medical profession disinclined to promote specialisation in experiment.[14] The turning point came in 1870, when Michael Foster moved from University College London to become first praelector in physiology at Trinity College Cambridge, where he would found the research school that was to dominate British physiology for the next thirty years. In the same year, the Royal College of Surgeons made physiology an examination requirement, probably under the influence of T.H. Huxley.[15] Any proper medical school now had to build a physiological laboratory. The early 1870s thus saw a big increase in animal experiments, and within five years the antivivisectionists had secured a Royal Commission on the subject. A year after that Parliament passed the Cruelty to Animals Act, which regulated research in Britain for more than a century. But although the Act marked a legislative endpoint, agitation continued into the 1890s and beyond. The entire episode offers important evidence about the kinds of feelings toward science which were abroad in Victorian Britain. And we can see the lasting impression of images informed by ideas about vivisection on the evolving stereotype of the scientist.

French situates the Victorian antivivisection movement in a time when the medical profession was working out how to deal with a new group of practitioners of experimental medicine, and with changes in clinical practice indicated by the results of new research. These two changes were part of a reconstitution of medicine on a scientific basis, but were also one element in a much wider debate about the place of science in late Victorian culture.

As ever, then, concerns about the specifics of scientific and medical practice were also informed by anxieties about the rise of science as a powerful social institution. The changes in medicine evoked heated opposition from sections of the public and of the profession. The opposition crystallised around movements against compulsory vaccination, against the Contagious Diseases Acts, and against vivisection. As French puts it:

all three appealed to the same kinds of fears of, and hostilities toward, science and medicine . . . [and] saw themselves arousing the moral instincts of laymen against an arrogant coalition of scientists, medical men, and legislators, who were blindly following the dictates of technique into an ethical cul de sac, where beneficent ends failed to justify horrid and repugnant means.[16]

What the three also had in common was that they all involved threats to the body. Compulsory vaccination meant state-enforced injection of foreign matter directly into the bodies of ordinary citizens.[17] The threat from vivisection was more indirect, but there were obvious links between what scientists were doing to animal bodies and what they might do to humans. So while the controversy over animal experiments was used to air a wide-ranging set of misgivings about science, these misgivings were brought into focus by the issue of what was permissible in biology.

The formal opposition drew largely on religious, moral or spiritual critiques of the scientific approach, while the physiologists emphasised its practical benefits. The legislative issue was the claim that needless cruelty was being inflicted on animals. As James Turner puts it, 'Not torture per se, but calm, calculating torture made vivisection the most heinous of crimes.'[18] French stresses that the treatment of animals in the laboratory came to symbolise the spiritual bankruptcy of science. The vivisectionist physiologists treated animals as mere experimental material; their opponents treated them as individuals capable of suffering.

The defenders of animals were also concerned with what kind of people the physiologists must be. As the romantic critique of science had long suggested, those who acquired knowledge by objectionable means would make bad use of it. The concern with animals thus fed back into a concern about human beings. If the antivivisectionists believed that animals had moral standing, they also feared that the physiologists would reduce people to mere biological systems, devoid of individuality or moral stature. They would be true to the spirit of La Mettrie's *L'Homme machine*, and would extend the anatomists' – and Victor Frankenstein's – attitude toward dead bodies to the living.

The renewed force of this critique was felt most strongly in fiction.[19] In the political arena, the abolitionists' campaign foundered because they broadened their attack from experimental physiology to the whole medical profession. The ensuing closing of ranks ensured that the new-found political power and prestige of the profession was enough to carry the day. The 1876 Act was framed, as French remarks, more to protect research than to restrict it. The promise of benefits to come helped quash misgivings about the costs of experiment. The antivivisectionist movement remained confined to small voluntary associations. Yet the public exposure of vivisection laid bare, and reinforced, deep reservations about the scientific enterprise, and especially about the development of biology, which still endure. They highlighted what an antivivisectionist polemic of a century later called *The Dark Face of Science*.[20]

This 'dark face' was, of course, already evident in literary portrayals of science and scientists, but the ambivalence toward science which the developing fictional stereotypes embodied was heightened in the last decades of the nineteenth century. The history of the British antivivisection movement implies a simultaneous endorsement of the therapeutic goals of the experimentalists and a repudiation of their methods.[21] But this was only part of a stronger contradiction in feelings about experimental biology.

On one hand, this was a time of celebration of medical science. Some fictional characters reflect this. Haynes describes how 'the realistic portrayals of nineteenth-century medical researchers are, almost without exception, complimentary to the point of eulogy':[22] Jenner's vaccination and Lister's antisepsis impressed almost everyone. Later, the first great practical successes of bacteriology were lauded as the work of scientific heroes, and Pasteur and Koch were viewed as universal benefactors. This was true in the US as well as Europe where, 'By the late nineteenth century . . . scientific discovery (almost all of it, of course European, but followed avidly by Americans), was touching the secrets and, therefore, ultimately control of even life and death. The germ theory of disease . . . brought with it more hope for the utter conquest of nature by man than any other theory or discovery of the day.'[23] Other advances in medical science were also enthusiastically received. They would lead to a stronger and healthier labour force, a more efficient army, more productive livestock, a longer life for all. In some ways, biomedicine came to stand for science in its entirety, as survey evidence suggests it may still do today.[24] The compelling interest of disease and death was for the first time united with a genuine practical promise. Most of the promises were as yet unfulfilled, to be sure, but the sense of crossing an historic threshold was widely shared.[25] In some ways, the effectiveness of the scientists' response to antivivisectionist agitation reinforced this image of research. James Turner, weighing up the debate at this time in Britain and the US, suggests that the vivisection controversy was as important as an opportunity to broadcast the new hopes of science as to express reservations about the paths research was taking.[26]

But if real scientists were mostly heroes, fictional scientists were quite another sort. The new-found powers claimed for science were not only cause for celebration. The conquest of nature, and control of life and death, were to be feared as much as welcomed, as a host of fictions already suggested. I have already described the endurance of *Frankenstein* in the nineteenth century, in versions in which the redeeming features of Victor and his creation have been pared away. The literary critic Milton Millhauser concludes a survey of the period with the suggestion that Stevenson's Dr Jekyll is the only other nineteenth-century scientific character of any lasting impact.[27] The Scotsman's brief parable of a fatally divided self, published to wide acclaim in 1886, has many similarities with *Frankenstein*, both in its origins in its author's dream-life and its extended after-life in multiple stage and film versions. But Stevenson's science is mere hocus-pocus, an unspecified 'drug' which frees the dark underside of Dr Henry Jekyll's personality. Even so, as well as its own mythic power, the story is notable for Jekyll's hatred of the body.[28] And Stevenson helped build the Gothic atmosphere of his story by deliberately evoking associations with dissection and vivisection. Dr Jekyll is not himself a vivisector, but his laboratory was once 'the laboratory or dissecting rooms' of a celebrated surgeon. The fact that it comes complete with a disguised entrance raises the question of what, exactly, the previous owner dissected there. Stevenson wrote in the city of Burke and Hare, so he knew very well what such entrances implied.

Henry Jekyll was one of the last of an important line of nineteenth-century scientists whose experiments were perverted or destructive. For Millhauser, the composite scientist in fiction was 'many unsavoury things in one person; wizard, alchemist, atheist, vivisectionist, poisoner'. He was Aylmer in Nathaniel Hawthorne's story 'The Birthmark' in 1843, Rappaccini in the same author's 'Rappaccini's Daughter' the following year, or the unnamed scientist in Jack London's turn-of-the-century story 'A Thousand Deaths'.[29] In London's story, the narrator is rescued from drowning at sea by a man whom he soon recognises as his long-lost father. The latter is a scientist who, like Victor Frankenstein, has 'abandoned the musty charms of antiquity and succumbed to the more fascinating ones embraced by biology'. As a result of his studies in 'physiology . . . organic chemistry, pathology, toxicology . . .' he succeeds in finding a means to revive the newly dead, by reversing the 'coagulation of certain elements and compounds in the protoplasm'. He verifies his theories by repeatedly killing his son and resuscitating him after longer and longer periods. His initially hapless victim finds his father 'the most abnormal specimen of cold-blooded cruelty I have ever seen'. In the end, the son masters the laboratory, kills his father, and we leave him preparing to carry on where his mentor left off. This kind of experiment dehumanises both the experimenter and his subject.

Millhauser suggests that the durability of the negative stereotype is bound up with the unfamiliarity of science and scientists, the threat to religious orthodoxy from the new findings of the physical, geological and biological sciences, and the doubts raised by the new claims of science in education. But he is also clear that the fictional expression of these misgivings centred on experiments in biology for much the same reason that positive images of science became identified with scientific medicine:

The occasional striking theory of a Newton, Lyell or Darwin was all very interesting, but it touched on men's lives less intimately and immediately than business, politics or even social aspiration and manoeuvring . . . Until quite late . . . most of the nineteenth century scientists were portrayed as concerned with medical or crudely biological experiments . . . It was *as a threat to his own body* – poison, narcosis, disfigurement, pain – that a layman most readily recognised the significance of science.[30]

In relation to vivisection, one kind of threat to the body was particularly felt by women. Coral Lansbury's illuminating study of antivivisection argues strongly that women's experience of medical practice, and their broader awareness of sexual subordination, formed an important subtext to the public debate about animal experiments. In her view, 'women were the most fervent supporters of antivivisection, not simply for reasons of humanity, but because the vivisected animal stood for vivisected woman: the woman strapped to the gynaecologist's table; the woman strapped and bound in the pornographic fiction of the period.'[31] Although Lansbury regards these influences as 'subterranean', it is not hard to find such links being directly drawn at the time. Turgenev's *Fathers and Sons* of 1861 centres on the life of a medical man committed to the new science. His first response on meeting the woman he loves is: 'what a magnificent body! Shouldn't I like to see it on the dissecting table!'[32] The abstract Baconian rhetoric of penetrating a feminised nature, which finds echoes in Mary Shelley, was now being reimposed on women by a male-dominated biomedical project.

In general, then, vivisection was a key influence on the way in which, for nineteenth-century novelists and their readers, science became 'the range of interests possible to an unusually ambitious, eccentric or depraved physician'.[33] Or perhaps he would not even have the saving grace of being a physician. Frankenstein began with a vision of ridding the human race of disease, but with the linked advent of professional physiology and vivisection the existing negative image underwent a significant change, even as it was re-emphasised by antivivisectionist propaganda. If lay suspicion of experimental medicine had formerly been partly allayed by the association with the doctor's care, specialisation meant that the two roles were growing more separate.

The antivivisection movement spawned a number of stories and novels which harped on the unemotional, unfeeling nature of the stereotypical physiologist. French offers Conan Doyle's 'A Physiologist's Wife' as a typical example of the public image of the morbidly curious physiologist. In this story, originally published in *Blackwood's Magazine* in 1890, the scientist makes a mental note to investigate the phenomenon of lachrymal secretion when his wife cries on his shoulder.[34] Some argued that this attitude would take over the whole medical profession, as in Lewis Carroll's (Charles Dodgson) vision of a day when 'successive generations of medical students, trained from their earliest years in the repression of all human sympathies, shall have developed a new and more hideous Frankenstein – a soulless being to whom science shall be all in all'.[35] But the association with medicine grew weaker as time passed. The image of the vivisector was incorporated into the composite image of the biological scientist, who now began to assume a separate identity.[36]

It was the fictional biologist who passed into the next century as the unfeeling obsessive. Jack London's experimenter is clearly identified as a scientist, not a doctor, and one of the fears which finally induces his son to fight back is 'the increasing predilection [my father] was beginning to portray towards vivisection'. He knows this because he wakes after one resuscitation with scars on his chest. The story speaks directly to the fear that the vivisector's attitude to animals extends to people. It was a logical corollary of the fears evoked by the anatomy murders. Contemporary scientists, like anatomists since the Renaissance, were prepared to dissect dead bodies. They were prepared to use bodies which had been killed for them. Why should they not go one better, and dissect a human body which was still alive? The answer was clear to some. As Lansbury suggests, 'the working class firmly believed that surgeons were vivisecting dogs, cats and rabbits because they could not vivisect human beings. When the latter were available, they would be used.'[37]

The influence of such fears can also be seen in 1888 in reactions to the murders of Jack the Ripper, who like Mr Hyde emerged from the fog-bound city streets to perform dreadful deeds under the noses of the authorities. It was widely believed that the murderer was a surgeon, and similar suspicions attached to other members of the medical profession. Martin Tropp records how at a time when: 'a coroner solemnly announced the absurd "fact" that an American doctor was paying for uteruses to supply with each issue of a new treatise on female disorders, medical men became suspects, shadowed by plainclothesmen, hauled in at the merest pretext.'[38]

A similar fear of human dissection, though it proves misplaced, appears in what is undoubtedly the most famous story to incorporate vivisection: Wells's *The Island of Dr Moreau*, first published in 1896. Vivisection is an essential part of the scientific rationale of the story. The plot hinges on the efforts of Moreau, a scientist ahead of his time, to create human-like beings from animals by extreme refinement of the vivisector's technique. In his other early scientific romances, all dating from the few years around the turn of the century, Wells made little pretence that the science behind *The Time Machine* or *The Invisible Man* was any more than hand-waving. Yet the first edition of *Moreau* carried a note – omitted from later editions – assuring readers that 'there can be no denying that whatever amount of scientific credulity attaches to the detail of this story, the manufacture of monsters – and perhaps even of quasi-human monsters – is within the possibilities of vivisection'.[39] The point was reinforced for contemporary readers by the fact that Wells had already explored the same ideas when writing as a scientific journalist. Much of the argument of Chapter 14 of the book, 'Dr Moreau Explains', had appeared the previous year in the *Saturday Review* under the title 'The limits of individual plasticity'.[40] In that chapter, Moreau's 'physiological lecture' as the narrator, Prendrick, calls it, the renegade surgeon-scientist emphasises how he is simply building on established possibilities of the technique:

> You forget all that a skilled vivisector can do with living things . . . For my own part I'm puzzled why the things I have done have not been done before. Small efforts of course have been made – amputation, tongue-cutting, excisions. Of course you know a squint may be induced or cured by surgery? Then in the case of excisions you have all kinds of secondary changes, pigmentary disturbances, modifications of the passions, alterations in the secretion of fatty tissue.[41]

He regales the appalled Prendrick with a long list of further examples, and concludes that 'it is a possible thing to transplant tissue from one part of an animal to another, or from one animal to another, to alter its chemical reactions and methods of growth, to modify the articulations of its limbs, and indeed to change it in its most intimate structure' (p. 78). After this manifesto for an experimental programme in physiology, he marvels that 'I was the first man to take up this question armed with antiseptic surgery, and with a really scientific knowledge of the laws of growth' (p. 79).

The political background, as well as the scientific detail, is authentic. Prendrick, when first cast up on Moreau's island, recognises the Doctor when he recalls an exposé of his experiments in England: 'The Moreau Horrors . . . I saw it in red lettering on a little buff-coloured pamphlet that to read made one shiver and creep . . . On the day of its publication, a wretched dog, flayed and otherwise mutilated, escaped from Moreau's house' (pp. 33–4). The perpetrator, 'howled out of the country', retreats to his island to continue his work unhindered. The British laboratory, isolated though it is, is not removed far enough from civilised restraints for the real possibilities of science to be developed. On the island, Moreau can develop his technique to the limit. The grotesque results of his labours, and those of his assistant, a drunken doctor, are the beast people who populate the island and flesh out Wells's evolutionary parable.

It is difficult, reading Wells's darkest novel, to remember that the author was, most of the time, a technological optimist. Mary Shelley encoded her ambivalences about science, knowledge and male power within her novel. In *Moreau*, Wells gave free rein to one side of his feelings about scientific possibilities and the limits of rationalism, possibilities for which he was more often a tireless advocate. Nor was he opposed to vivisection. He knew his biology of course, learned from T.H. Huxley at South Kensington, and his own textbook of biology is described by Parrinder as virtually a dissection manual.[42] In later life he wrote a defence of vivisection for the *Sunday Express* which drew a rebuttal from G.B. Shaw. Nevertheless, the representation of experiment as actual manipulation of organisms which his fictional use of vivisection permitted was crucial to his literary intent – to blur the distinction between men and animals. When Moreau is killed by one of his creations, his pathetic half-human creatures revert to an animal existence. The reversion of the Beast People, and the repatriated Prendrick's subsequent inability to distinguish between real humans and his memories of the creatures on the island, is the main thrust of Wells's post-Darwinian rendering of Swift.

The later Wells, who disparaged all the scientific romances, was to describe *Moreau* as 'an exercise in youthful blasphemy', referring to Moreau's figuring as a ruthless God of (un)natural selection. Contemporary reaction focused more on the *horror* of the book. *The Times* called it 'perverse', 'loathsome' and 'repulsive', and others added that it was degrading and nauseating.[43] The *Critic* warned that it might prove dangerous 'even for physicians and scientific men – especially those of an experimental turn of mind'. As Baldick observes, *Moreau* was a recasting of *Frankenstein* distinguished by becoming 'the first member of this class to stir the same outrage and disgust among its first readers as *Frankenstein* provoked'.[44]

At least two reviewers read the book as an attack on vivisection, R.H. Hutton in the *Spectator* praising it on that score while Sir Peter Chalmers Mitchell, the zoologist and Wells's colleague on the *Saturday Review* confessed to the 'frankest dismay' at the portrayal of Moreau as a 'cliché from the pages of an antivivisectionist pamphlet.'[45] He criticised Wells for harping on the pain suffered by the animals, ignoring the use of anaesthetics, and scaring the public with tales of monsters created by vivisection. Raknem observes that this criticism was apparently 'a huge advertisement' for the book – not the last time a scientist's reaction to a piece of fiction would achieve the opposite of the desired effect.[46]

Moreau is the most extreme example of a pessimistic strain in the fiction which made Wells's name as he made a successful switch from journalism to literature. It was produced in a remarkable creative burst which Wells the serious novelist later belittled, but which produced his most lasting fictions. *Moreau* appeared the year after the author made his reputation with *The Time Machine*, and shortly before *The Invisible Man*, which also centres on the ruthless exploitation of a scientific discovery. Of the three, it is the most deeply rooted in Wells's own scientific training, and in knowledge of contemporary research. This is one reason why it did more to reinforce and enrich pre-existing scientific stereotypes than his other books. Chalmers Mitchell did well to protest against the identification of the new experimental biology with the project of creating monsters. The picture Wells draws of Moreau, orchestrating his cruel parody of natural selection, is deliberately that of the modern scientist, stripped of all fellow-feeling for other creatures, and of ethical qualms. As Hillegas suggests, 'Moreau is a far more sinister creature than the medieval Faust or Mary Shelley's Frankenstein. He has [T.H.] Huxley's command of the powerful scientific method; but he does not have Huxley's controlling humanity.'[47] This dehumanisation is the direct result of Moreau's research. He tells Prendrick that he has devoted his life to 'the study of the plasticity of living forms'. Like Victor Frankenstein, the imperatives of science led him on and on:

> I asked a question, devised some method of getting an answer, and got –
> a fresh question. Was this possible, or that possible? You cannot imagine
> what this means to an investigator, what an intellectual passion grows upon
> him. You cannot imagine the strange colourless delight of these intellec-
> tual desires. The thing before you is no longer an animal, a fellow creature,
> but a problem . . . The study of Nature makes a man at last as remorseless
> as Nature. (p. 70)

As well as its reference to the Darwinian view of the struggle for survival, this passage is a startlingly close echo of the words of Bernard's *Introduction to the Study of Experimental Medicine*:

The physiologist is not an ordinary man; he is a scientist, possessed and absorbed by the scientific idea that he pursues. He doesn't hear the cries of the animals, he does not see their flowing blood, he sees nothing but his idea, and is aware of nothing but organisms which conceal from him the problems he is wishing to resolve.[48]

Unlike *Frankenstein*, where the deliberate vagueness about the manner of life creation has encouraged substitution of more recent science in later reworkings, it is the very specificity of Moreau's technique that has perhaps helped the story endure. Now that modern biology was set on its path, Wells was able to single out a practice which for many still stands for what goes on in the laboratory. Certainly, critics are quick to claim the contemporary relevance of the novel.[49]

More significant, I think, is its contribution to the composite myth of the biologist. Moreau, a much more sharply drawn figure than Victor Frankenstein, appears to have none of the humanitarian motives which redeem Victor's project when it begins. He simply wants to show that he can refashion living things according to his own design, and admits to Prendrick that he chose the human form as a model 'by chance'.

Moreau is as obsessed as Frankenstein, but far more sinister in his detachment. 'I looked at him and saw but a white-faced white-haired man, with calm eyes. Save for his serenity, the touch almost of beauty that resulted from his set tranquillity, and from his magnificent build, he might have passed muster among a hundred other comfortable old gentlemen' (p. 87). One consequence is that, while Victor almost kills himself with the effort of making one creature, Moreau pursues a lengthy experimental programme, an attempt to accelerate evolution: 'each time I dip a living creature into the bath of burning pain, I say, This time I will burn out all the animal, this time I will make a rational creature of my own. After all, what is ten years? Man has been a hundred thousand in the making' (p. 86).

The creation of a whole family of monsters yields a range of images which linger, rather than that of a single being. The Beast People, a memorable blend of creatures from a medieval bestiary and beings of Wells's own invention, harbour a fair proportion of the cinematic monsters of the next century. They include, as an aside, 'a limbless thing with a horrible face that writhed along the ground'. Of this 'thing', which escapes and causes havoc before it is finally killed, Moreau says, in a scene which now seems very familiar: 'It only got loose by accident – I never meant it to get away. It wasn't finished. It was purely an experiment' (p. 85).

Like *Frankenstein*, which it follows so closely in many respects, Wells's short novel is open to a host of interpretations, and it has some of the qualities of myth.[50] As with Mary Shelley at the beginning of the century, the power of Wells's imagination created a unique and horrific vision from contemporary scientific and political ingredients; a vision which, largely bereft of its original philosophical import, has entered popular culture. Once again, the elements which survive many refractions of the original image include the identification of the laboratory scientist as a biologist. Also enduring are the features shared with Mary Shelley's novel. Both 'depict the takeover of a natural female function by crazed male science',[51] and both share the theme of escape – the scientist's escape from moral constraints and the escape of the forces unleashed by his breach of old boundaries. But now the breach is achieved by 'pure' experiment.

Improving the species

There was a further set of developments colouring attitudes to the rise of biology, developments concerned not so much with experimental science and the increase in animal experiments but more directly with human breeding. Wells's dark satire was one possible response to Darwinian theory. Another was the enthusiasm for eugenic ideas, prompted both by dreams of human improvement and fears of racial degeneration. Their lasting influence on attitudes to biology as a potential science of social control also has its origins during the second half of the nineteenth century.

The eugenics movement in Britain, a kind of secular religion which reached its high point of influence in the years just before World War I, was the product of a variety of hereditarian currents dating from the late nineteenth century, and of certain features of Edwardian society. These included the rise of organised labour and a fear that some kind of national degeneration would lead to an inability to police the Empire. Eugenic theory, first laid down by Darwin's polymathic cousin Francis Galton, was ostensibly grounded in biological science, and brought biology to bear on fundamental human concerns.[52] The idea of racial decline was consonant with a vulgar Darwinism – and problems which had been disclosed by science could only be solved, according to the eugenists, by application of scientific principles to social organisation. The supporters of eugenics drew heavily on the high prestige of science as an agent of progress and social change.[53]

Yet, there was a difference, in degree if not in kind, between the exploitation of the social prestige of science by the eugenists and more general contemporary endorsements of scientific progress. Here, the authority of science was being invoked in an explicit attempt to impose a scheme devised by one section of society on to another. In Mackenzie's words, eugenics in Britain was an 'ideology of the professional middle class'.[54] And although in principle they sought a general uplift in quality, in practice their attentions were directed downward. It was generally accepted that too little was known to pursue a selective breeding programme aimed at positive eugenics. So eugenic schemes concentrated on eliminating the unfit – on negative eugenics. Since social status was equated with biological worth, negative eugenics focused on the lower classes.

Unlike previous defences of the status quo on the basis of scientific theory, the eugenists' position was interventionist. The essentially *laissez-faire* doctrines of social Darwinism had ocasionally given rise to calls for sterilisation of the feeble-minded,[55] but it was not until the rise of an organised eugenics movement that such suggestions commanded widespread support. Those designated as 'unfit' were now threatened with enforced segregation, sterilisation or even, on occasion, with the lethal chamber.[56] The eugenists' efforts in Britain were not usually directed against the working class as a whole, but against the urban poor, the feeble-minded, drunkards, epileptics – the 'residuum'. According to Mackenzie,

> All eugenists were agreed that manual workers were socially necessary. What they wanted was to improve the discipline, physique and intelligence of the working class by eradicating the 'lowest' elements of it. The eugenists attempted to draw a line between socially useful and socially dangerous elements of the lower orders.[57]

What effect did this have on attitudes to science? It is hard to give a clear answer, even now. It seems that working-class opposition to eugenics was muted in Britain. According to Searle, trade union leaders mostly ignored eugenics, or were even unaware that it existed. Opposition came, rather, from Roman Catholics and 'a certain kind of individualist liberal'.[58] Nevertheless, this move to extend the modern project to the regulation of reproduction must have had some impact on those who would be subject to regulation.

Farrell suggests that eugenics may have had some such effect, though only obliquely. She acknowledges that by the end of the nineteenth century 'science' conveyed 'quite different meanings to different people', but nevertheless affirms that it 'was everywhere viewed as the working man's friend', because of its contributions to everyday comfort and national prestige. However, the main evidence she cites for this conclusion is an unpublished study of newspaper reporting of science in a popular newspaper in the 1870s, well before the peak of the eugenics movement.[59]

More suggestive are the findings from cultural historian Peter Broks' extensive survey of science in mass circulation British magazines around the turn of the century. In reviewing the contents of magazines like *Pearson's Weekly* and *Tit-Bits*, with circulations in the hundreds of thousands, he finds a much more ambivalent image of science.

In fiction, these magazines certainly carried on the familiar tradition: 'Almost to a man . . . fictional scientists were portrayed as, at best, unemotional and detached, and, at worst, inhuman and insane. They actively connived at the deaths of others, failed to see the dangerous implications of their work, or were indifferent to the suffering that their research might cause.'[60] Among many examples, these fictional scientists include a bacteriologist with an ultra-eugenic scheme 'to kill off the working population to solve the problem of poverty'. Contrast this with the real scientists, who were glowingly described as heroes of intellect and paragons of dedication. Although Broks detects a tendency toward disillusion with the fruits of scientific progress as the Victorian era gives way to the Edwardian, the shift is part of the climate which fostered the eugenics movement, rather than a reaction to it. It is evident in a growing concern with racial deterioration and the burden of the 'unfit'.

It seems, then, that it is retrospective views of eugenics that have played the main role in promoting a political analysis of the development and uses of biology, rather than arguments or reactions registered when eugenic sentiments first became a force to be reckoned with. In particular, of course, the later eugenic practices of the Third Reich cast the earlier eugenics movement in a much more negative light, and had an enduring impact on public discussion of potential uses of more recent biology. But it is still possible to suggest that this response to eugenics has roots in the Edwardian era, and we can turn to imaginative literature in the US to support the suggestion. The eugenics movement in the US was different in several respects from that in Britain, although the British movement served as the main inspiration for

American eugenists' doctrine.[61] The most important difference was that eugenic sentiment in America was mainly directed against immigrant groups, rather than a native underclass. In terms of popular response, the difference is to be found in the fiction of the period.

A good deal of what is now termed science fiction was published around the turn of the century, especially in the US. With an increasing number of stories depicting science and scientists, the almost exclusive concentration on medical investigators began to weaken, and the optimism of the period largely overcame the classically negative image of the scientist – even though Dr Moreau was still fresh in the public mind.[62]

But if the scientist, or more usually the inventor, was now more often seen as the hero of the tale, albeit often also an innocent or a buffoon, biological scientists were still more likely to be viewed with suspicion. And the ground for that suspicion now also included their eugenic sympathies. While the more typical stories were concerned with discovery of a lost race, voyages to the moon or Mars, or celebrations of technology, the explicitly biological stories often had a darker hue. For example, all the stories cited in Clareson's survey of American science fiction between 1890 and 1915 which are concerned with biological themes are critical of science. They include H.O. Cummins's 'The Man who Made a Man' from 1902,[63] a short story reworking of *Frankenstein*, as was W.G. Morrow's 'The Monster Maker' of 1897.

More complex was Will Harben's novel *The Land of the Changing Sun*, published in 1904, which is summarised by Clareson as about 'a lost race which intentionally colonized a great subterranean cavern near the Arctic and developed a man-made electrical sun to heat and light it. The race has a kind of obsession regarding eugenics; Harben bitterly attacks the medical profession for advocating a eugenics program for the world.'[64] Also cavernous was J.U. Lloyd's *Etidorhpa* of 1895, which depicts a journey into the earth's interior by a man who has betrayed the secrets of alchemy and been punished with eternal life. According to Clareson it contains 'one of the most violent attacks on science, especially biology, of the period'.[65]

But it is an even less well-known novel which yields the most elaborate response to eugenic ideas. *The Messiah of the Cylinder*, by Victor Emanuel, was published in 1917, and intended as a riposte to the scientific optimism of H.G. Wells's *When the Sleeper Wakes*.[66] The book opens in 1917, with a description of a scientist working at the 'Biological Institute'. This man, Lazaroff, is not only a racist eugenist, but also a vivisector. He is soon transferred, by an experiment in suspended animation, to a decadent twenty-first-century capitalist world, where he is hailed as a messiah by the followers of a cult which has grown up around his sleeping figure in its translucent cylinder. Then we see what science unleashed can do. Lazaroff exploits his new following to lead a revolution and is soon installed as co-dictator of a regimented scientific state. The population is scientifically ranked into eugenic classes, and only the highest classes are encouraged to reproduce freely. The lowest class – the morons – are not only prohibited from repro-ducing, but are also forced to serve as fodder for biological experiments on living subjects. Lazaroff, like Moreau, becomes embroiled in a vast pro-gramme of experimental surgery on human beings, in the hope of improving the race and eventually making men immortal. He is willing to sacrifice unlimited numbers of the lower classes to achieve this end.

The significance of Emanuel's dystopia lies in its combination of ideas rather than its contemporary impact. Although at least one popular magazine recommended the book to readers who had enjoyed Wells, it attracted few other responses.[67] Nevertheless, it is striking to see such a well-developed vision of the consequences of contemporary biology combined with absolute political power – and of a marriage of eugenic goals and modern experimental methods. The addition of a critique of eugenics as a vehicle for social oppression to individual fears of application of experimental techniques to human beings is an important development of previous images. Once again, a book embodying powerful negative feelings about science adopted aspects of contemporary biology to symbolise the human impact of scientific developments.[68]

Emanuel's 'Biological Institute' was a fictional establishment, but was depicted in the knowledge that real biological institutes were now flourishing in the US and Europe. By the end of the first decade of the twentieth century, the rise of experimental biology had brought to prominence a new way of approaching the study of living things, scientifically, institutionally and pub-licly. And as the century turned, it appeared to some that experiment was on the brink of achieving some of the promises implied by writers of fiction. The next chapter explains how this came about, and how the public responded.

Chapter Four

Creating Life in the Laboratory

By the turn of this century, the practice and ambitions of biological research were much closer to their modern form, and the responses evoked by the reported work of biologists were also in some ways more similar to those we know today. In this chapter, I examine a remarkable episode in the public history of biology, when it seemed to many that experimentalists really were on the verge of creating life in the laboratory. It seems hard to credit now that this claim was being made – by reputable scientists – and believed, before World War I. But the state of knowledge, the philosophical and scientific ambitions of some researchers and the newspaper treatment of science at the time combined to produce this remarkable expectation. The reactions prefigure those which we find later this century as claims to be able to manipulate life acquire a stronger basis in reality. But first we must look at how the notion that life might be created was built up, mainly through the public pronouncements of two very prominent, but contrasting figures, Jaques Loeb and Alexis Carrel.

From spontaneous generation to creating life

As biology became established as an experimental science, the unifying power of three great scientific developments of the nineteenth century – the cell theory, Darwinian evolution, and thermodynamics – fostered a conviction that the fundamental problems of living matter were finally accessible to study. And while spontaneous generation as a contemporary phenomenon had been discounted by the vast majority of scientists after Pasteur's rebuttals of the 1860s, the materialist Darwinian scheme drew attention to the possibility of a spontaneous generation at the point of origin. Immediately the question arose: if it could happen once, could people make it happen again?

For some optimistic commentators, as yet unaware of the complexities that would be unveiled inside living things in the coming decades, the answer at the dawn of the new century was: why not? As this new confidence was broadcast outside the community of biologists, the question in popular discussion became not *would* biologists create life, but *when?*

This optimism can be seen in an article on 'Some unsolved scientific problems' in *Harper's Magazine* in 1900. In this, the last of a series which reviewed in glowing terms the achievements of nineteenth-century science, the view turns to the future. For biology: 'It seems well within the range of scientific explanation that the laboratory worker of the future will learn how so to duplicate telluric [i.e. planetary] conditions that the play of universal forces will build living matter out of the inorganic in the laboratory.'[1]

The warrant for this assertion is Darwinian:

To the timid reasoner that assumption of possibilities [that life may be created] may seem startling. But assuredly it is not more so than seemed, a century ago, the assumption that man has evolved, through the agency of 'natural laws' only, from the lowest organisms. Yet the timidity of that elder day has been obliged by the progress of our century to adapt its conceptions to that assured sequence of events. And some day, in all probability, the timidity of today will be obliged to take that final logical step which today's knowledge foreshadows as a future if not a present necessity.

The life created would presumably have been protoplasm, in the sense that emerged in the late nineteenth century. The concept grew out of the cell theory of the 1840s, and the developed notion of protoplasm as a universal ground substance of living things was popularised by T.H. Huxley in his famous address of 1868 in which he identified this substance as 'The physical basis of life'. The address was published in the London *Fortnightly Review* and the New York *World*, so that the materialist definition of protoplasm reached a wide audience.[2]

For Huxley, the chemical make-up of protoplasm was essentially identical in all living things: it was 'built up of ordinary matter and again resolved into ordinary matter when its work is done'. On this view, the question of whether living matter could be produced in the laboratory reduced to the possibility or otherwise of duplicating protoplasm. At first, few entertained high hopes. The identification of protoplasm as a material substance was of course contentious, and even those who accepted it felt that it must be extraordinarily complicated. According to Gerry Geison, 'Probably the most popular theory about the nature of life throughout the rest of the nineteenth century was that it resided in a material substance, as Huxley proclaimed, but that this substance when alive was of a composition so uniquely complex that it was unlike ordinary matter.'[3]

This complexity was overlooked by some, and seen as a challenge by others, especially by a younger generation of materialist biologists. When occasional claims surfaced that protoplasm could be created, claims which required a rather simple concept of what it might consist of, they were widely reported. An early example is the work of Butschli in the late 1890s. This young researcher, professor of zoology and palaeontology at Heidelberg, was a mechanist who believed that many of the properties of protoplasm could be explained in physico-chemical terms. He conducted a series of experiments in which potassium carbonate was mixed with various oils and the microscopic resemblance of the saponified foam to living protoplasm recorded. He published a major monograph on these experiments, illustrated with hundreds of photomicrographs, and accounts of the work found their way into the press. They were the first of many. In 1911, *Scientific American* lamented that

> Some fifteen years ago . . . these experiments were hailed by certain journalists as successful attempts to create artificial life . . . How hard error does die is suggested by the fact that these experiments have been described over and over again in all grades of publication as a method for making 'artificial protoplasm' until this very year.[4]

In fact, these experiments became part of a constellation of studies which were identified with the creation of life, each of them drawing plausibility from the others. In particular the idea was associated with US-based figures, Jacques Loeb and Alexis Carrel, both of them recognisably 'visible scientists' in the sense Rae Goodell has applied to more recent researchers with a high media profile.[5] Goodell suggests that her visible scientists, consistently successful at making news, are 'unique to contemporary media and their audiences'. I agree, in the sense that the political and personal goals of scientists who were in the public eye in the early 1900s, or the demands of the media, are unlikely to be the same as those of their present-day equivalents. However, the concept is a useful one, as it focuses attention on the conditions of visibility. What was it that distinguished the researchers whose work received extensive publicity from their comparatively 'invisible' colleagues?

The general level of science coverage in the press at this period was low. Only three American newspapers, for example, had expert science writers on retainer in 1907.[6] In spite of this, the idea of creating life commanded wide coverage, suggesting something of its hold on the popular imagination. Loeb's and Carrel's links with the idea came about in different ways, but this seems to have had little effect on the general attitude to their work.

The 'chemical creation of life' – Jacques Loeb and artificial parthenogenesis

Jacques Loeb was the first major scientist to be publicly linked with the possibility of life creation, and his work was used to lend plausibility to the idea over a long period. He was also frequently portrayed by his contemporaries, as he is by historians, as personifying the attitudes of the experimental biologists of the time. Scientifically, Loeb was an extreme case but, partly because of this, an influential one. He was an arch-mechanist, committed to an experimental programme which sought above all control over living things. His combination of experimental flair and a metaphysical commitment to materialism whose fervour outstripped that of any of his colleagues made him the best-known biologist in the United States in the early twentieth century.[7]

Loeb was a German Jew who brought a lengthy training in mechanistic physiology and the latest physico-chemical methods with him when he entered the USA in 1891, aged 22. He worked at Bryn Mawr and Chicago (institutions which also employed the other three chief apostles of the experimental method in biology at the time – Ross Harrison, E.B. Wilson, and Loeb's close friend T.H. Morgan).[8] It was at Chicago that he carried out the experiments which first earned him real notoriety: on the artificial fertilisation of sea-urchin and, later, frogs' eggs. This was work which derived from the general pattern of research in this period. The experimental and mechanistic approach to embryology was heavily influenced by the programme for *Entwicklungsmechanik* – developmental mechanics – elaborated by Wilhelm Roux in the 1890s. T.H. Morgan began his research career as an embryologist before turning to genetics, and Ross Harrison came to dominate the subject in the USA. The sea urchin was the experimental organism of choice for all these men, being readily available at the shoreline laboratories at Woods Hole and Pacific Grove, both modelled on the famous Naples research station in Europe.[9]

It was in fact Loeb's friend Morgan who found in 1899 that, every now and again, an unfertilised egg could be induced to divide in hypertonic sea-water. But while Morgan thought the phenomenon pathological, Loeb, with his burning desire to find mechanistic explanations for all natural phenomena, seized on Morgan's stray observation.[10] As a colleague later related, he

> immediately grasped the potentialities of such an observation and its immense value towards his general objective. The process of fertilization . . . was surely the most mysterious thing in the world. If he could strip the mystery from this and show that it was possible to imitate fertilization by some physical or chemical agent, a deadly blow would have been delivered against that sort of mysticism which revels at our ignorance of natural phenomena.[11]

When Loeb repeated and elaborated Morgan's findings, there was, in Reingold's words, a 'sensational journalistic flurry'.[12] As Pauly recounts in detail, 'artificial parthenogenesis entered popular consciousness with the new century', with the first newspaper headlines announcing the creation of life appearing toward the end of 1899.[13]

And the story endured well into the 1900s. The first reports were reproduced for more than a decade.[14] The interpretation put on the work was very much in line with Loeb's own predilections. It was tied in with an awareness of the growing power of experimental biology, and with the potentialities of that power. Among journalists, Loeb's experiments came to be habitually described as demonstrating the creation of life. They became one of the collection of pieces of research, which towards the end of this period were often treated together, that were taken as supporting the idea. This notion of artificial creation became a journalistic leitmotif, a way of encapsulating new expectations for biological science in a single phrase, and of evoking succinctly the spirit of the whole enterprise.

What role did Loeb himself play in this? Certainly, he was generally highly circumspect in his descriptions of actual experiments, and direct quotations from his bulletins are dry accounts of the bare scientific details. It was in headlines that the interpretive work is most obvious. Thus, the *New York Times* for 1 March 1905 headed an account of the work on artificial parthenogenesis,

'Chemical Creation of Life', a phrase closely paralleled by the headline in the *Chicago Tribune* the same morning: 'Creates Life by Chemistry'. On 14 September that year the *New York Times* came up with 'Near Explaining Life by Loeb's Experiments'. The *Tribune* told readers that 'Professor Jacques Loeb . . . has just made public another step in his work of producing life by artificial means'. Loeb's results, and the way they were reported, also set a precedent for coverage of work by others. In 1907, his observations on sea urchins were confirmed in Paris, and in 1911 extended to frogs by Bataillon in Dijon. To give just one report on each from the *New York Times*, the Paris work by Delage was described under the headline, 'Uses Chemicals to Produce Life' (25 September 1907); on 8 August 1911, under the heading 'Tadpoles generated with Electricity's aid', readers were told that 'success in experiments leading toward artificial generation of life is now claimed by a biologist of Dijon, Professor Bataillon'.

Needless to say, there were also much wilder reports, even in the *New York Times*. On 4 October 1910, the following item appeared in that paper under the heading, 'Chemical Human Embryo':

Vienna, Oct 3. – The Mexican consul in Trieste reports that Prof. Herrera, a Mexican scientist, has succeeded in forming a human embryo by chemical combination.

Fairly far-fetched, but as the writer remarked, 'recently the best known experiments in the line of the artificial generation of living embryos have been those of Jacques Loeb'.

It was obvious to more serious commentators that the phrase 'creating life' was an overstatement of what was done in Loeb's dishes, just as it is obvious today that *in vitro* fertilisation does not literally produce 'test-tube' babies. The *New York Times* leader writer took a sceptical view of Bataillon's work in 1911, declaring that 'to describe it, as some of the dispatches do, as "the artificial creation of life" is a manifest error and absurdity'.[15] After all,

The biologist had 'life' in being no less at the beginning than at the end of his research, and of its 'artificial production' there was not a trace. In other words, he did not make dead matter live, which would indeed be artificial, nor have any of his predecessors in this line of investigation ever done so.

Yet the same paper, reporting Loeb's successful repetition of Bataillon's work with frogs, told readers that 'Dr Jacques Loeb, of the Rockefeller Institute, has succeeded in demonstrating how life may be produced by artificial means'.[16]

Journalistic licence aside, it was Loeb's own attitude to the experiments on artificial experimentation which provided the most powerful legitimation for this interpretation, though he never actually gave it his unqualified support. His attitude to the public accounts of his work was highly equivocal: he disapproved of the wilder exaggerations, but he also wanted to promote his mechanistic programme for biology as strongly as possible.

Loeb shared the prevalent view that popularisation of science was illegitimate. Most of his contemporaries were uninterested in such things. Tobey relates how when the National Academy of Sciences proposed in 1916 that it might sponsor a new magazine of popular science, the idea had a cold reception from members. The great cytologist E.B. Wilson had 'a rather poor opinion of scientific men for running a popular journal'. Tobey comments that 'there was just the slightest intimation in Wilson's tone that it was beneath the dignity of scientists to explain their work to laymen'.[17] Loeb, true to this tradition, appears not to have enjoyed his notoriety. Several associates report his distress when his work became the focus of popular attention.[18] However, he did believe that life could be created, and he was not prepared to deny it. After all, it was an explicit part of his overarching programme for experimental biology. Sympathetic commentators like Wilhelm Roux wrote at the time that

> The political newspapers and popular science journals are publishing accounts of the artificial generation of life and exciting universal amazement among their readers. [however] . . . the element of amazement arises . . . mainly from the interpretation put upon the matter by the imagination of these accounts; the experimenters themselves speak with considerably greater caution.[19]

But this was wide of the mark. For against Loeb's professed distaste for publicity there was acting a more powerful force: his metaphysical convictions. While he did not endorse the idea that artificial fertilisation itself was equivalent to the creation of life, the idea, once raised, became for Loeb an important test of the mechanistic approach to biology. He took it as an implicit goal for future research, and suggested that it would be the final vindication of his views.

In 1910 Loeb acquired still higher visibility by joining the staff of the Rockefeller Institute for Medical Research in New York. Here he came to play a key role in an institution very much in the public eye. The Institute was extremely well endowed and the combination of Rockefeller's financial interest and the establishment's declared aims ensured that expectations were high. When it was incorporated in 1901, according to Ronald Tobey, the event was fully discussed in the New York newspapers, which saw it as an effort to remove America's deficiency in original research. 'Because of the reputation of European medical research institutes, such as the Pasteur Institute and Koch Institute, public esteem for the Rockefeller Institute was assured.'[20] Loeb's recruitment was an important step in the development of the Rockefeller research programme. He was appointed to head a new division of experimental biology, with a salary second only to that of the director of the Institute. It was a high-profile job, both within the scientific community and outside, and one he took full advantage of. As the microbiologist René Dubos, who came to the Rockefeller later, recalled, 'He soon became the most influential member of the Institute staff . . . By brazenly parading his mechanistic animus, he did more than anyone to foster the belief that the most effective approach to biology and medical research is through physics and chemistry.'[21]

One way this 'mechanistic animus' was expressed was in the interpretation he still encouraged of his most famous experiments. His most explicit association of the artificial fertilisation experiments with the creation of life came in his most famous essay, 'The mechanistic conception of life', the text of an address given at the first International Congress of Monists in Hamburg in 1911. It was later published in the *Popular Science Monthly*, and as the opening essay in a book of the same title.[22] It was the book which prompted the *New York Times* reviewer, Carl Snyder, to hail Loeb as 'undoubtedly one of the greatest experimental geniuses of whom we have any knowledge'.[23]

In the address, Loeb made his now customary claim that the 'complete physico-chemical explanation of fertilisation' was a paradigm for the demystification of biology: 'Considering the youth of biology, we have every right to hope that what has been accomplished in this problem [fertilisation] will occur in rapid succession in those problems which still today appear as riddles'.[24] He had already indicated what kind of riddles he meant: 'Nothing indicates . . . at present that the artificial production of living matter is beyond the possibilities of science . . . *We must either succeed in producing living matter artificially or we must find the reasons why this is impossible'* (emphasis added). For Loeb, obtaining control over biology was a step toward the rational organisation of society. But what came across to the casual newspaper reader was the potential for control of the new experimental biology, symbolised by the prospect of creating life artificially. This is underlined by a further selection from contemporary newspaper reports.

For the *San Francisco Chronicle*, 'the discoveries of the secrets of these two processes [sexual and artificial fertilisation] will mean the unlocking of much of what has always been a closed mystery to science'.[25] For the *New York Times*, 'the series of extraordinary discoveries recently made in New York came very near to the mystery of life'.[26] In Loeb's laboratory, according to a picture caption in *Cosmopolitan Magazine*, 'the tanks full of prisoners kept in running water, and the great microscopes standing in stiff, staring attitudes . . . suggest the spirit of investigation and of prying into Nature's secrets'.[27]

As a result of reports like these, Jacques Loeb's work was the most important source of speculation that life might be created in the laboratory. But this prospect combined with popular accounts of other research to produce a broader impression of the growing manipulative potential of biology. Another figure of almost equal importance in sustaining this impression was the flamboyant French surgeon Alexis Carrel, for a long while Loeb's colleague at the Rockefeller Institute. In press reports, the rather different work of these two men was often described in ways which suggested they disclosed separate aspects of the same endeavour. For this reason, I explore the sources of Carrel's 'visibility' before considering the reactions evoked by the researches of both men.

'The heart in the jar' — Alexis Carrel and the origins of transplant surgery and tissue culture

Experiments which could be represented as creating life were perhaps the central part of the image of a biology with a new manipulative potency which was built up in the early 1900s. But there was also other work which was used to reinforce the impression that laboratory workers were learning how to control life. The suggestion that the life-creation stories might be true lent credibility to claims about manipulation of existing life.

In this context, the work of Alexis Carrel is especially interesting. In accounts of his achievements, there is a persistent emphasis on control. The tangible surgical manipulations at which he excelled were easier to describe than chemical experimentation, and actual bodies rather than cells were easier to visualise. At the same time, though he himself appears to have had little interest in the matter, his work was used by others to support the claims surrounding life creation. The climate of expectation embodied in this idea led to greatly exaggerated accounts of Carrel's actual achievements appearing in the press.

Carrel, another recent immigrant to the US, was born in Lyons in 1873. After medical training in Lyons he started to experiment on blood vessel surgery, and began to develop the techniques and the dexterity which were the hallmark of his later work.[28] Philosophically, he was ostensibly Loeb's opponent. He testified to a miracle cure at Lourdes in 1903, and retained a lifelong interest in mysticism, clairvoyance and other psychic phenomena. But, like Loeb, he had a flair for publicity, perhaps unwitting at first. An early encounter with the press in connection with the Lourdes miracle helped persuade him to leave France, and in 1903 he began work in the Hull Laboratory of Physiology at the University of Chicago, attempting organ and limb transplants. Two years later, he was offered posts at both Johns Hopkins Medical School and the Rockefeller Institute, and he became head of the Department of Experimental Surgery at the Rockefeller in December 1906.

Like Loeb, he found himself in the public eye at the Rockefeller. The directors of the Institute were keen to publicise its work, and scarcely a month passed without a report of some new discovery or cure from the Rockefeller laboratories. G.W. Corner, in his largely uncritical history, remarks that 'the Institute and Flexner himself [the director] have been accused of capitalizing prematurely on some of the hopeful results of the early research'.[29] In 1903 the Institute had acquired the *Journal of Experimental Medicine* as an outlet for its own papers, and Corner relates how reporters, seeking more sensational material, would comb each issue. Carrel's work was one of their main finds. According to Corner, 'Of all the Institute's varied researches in the years from 1906 to World War One, Carrel's tissue culture work most vividly attracted public attention.' The reasons were twofold: 'Scientists saw in it an important new way of studying life processes; laymen were astonished and awed by the idea of living cells growing and multiplying in an incubator, and even attaining the semblance of immortality by long outliving the creatures from which they were explanted.'[30]

Carrel's contribution to tissue culture was opportunistic. The techniques were pioneered by Ross Harrison at Yale and quickly taken up by others. But they soon became identified with Carrel because he made the strongest claims about what he had achieved. As with Loeb's experiments on artificial ferti-lisation, the tissue culture work was firmly grounded in the mainstream of experimental biology of the time. The problem of maintaining cells of a multicellular organism outside the body had long perplexed cytologists. The solution was found by Harrison during experiments designed to resolve a long-standing dispute in embryology about the mode of propagation of nerve fibres. He excised a section of spinal cord from a chick embryo, suspended it in a clear drop of lymph on a slide – and was able to show the continuous growth of the nerve fibres.

Harrison's original paper appeared in 1907, and Carrel heard him lecture on the technique in New York in March 1908.[31] Immediately he sent one of his assistants to learn how it was done and when he returned to the Rockefeller the pair soon succeeded in culturing cells. They were one group among many:

> In the years following publication of Harrison's full paper in 1910, the method was enthusiastically adopted by workers throughout the world . . . A very large number of papers were published and by 1911 tissue culture was sufficiently well established for the American Association of Anatomists to devote a symposium to it at their annual meeting in Princeton.[32]

Yet it was the Frenchman who became identified with the technique, partly because of his technical flair, partly because of the Rockefeller connection, and partly because of his response to scientific criticism.

He soon progressed from culture of embryonic tissues to adult cells, and reported in October 1910 that his laboratory had successfully grown several types of adult tissue, including thyroid and kidney, in culture media. Carrel claimed to be able to detect structures characteristic of these organs under the microscope. In November 1910 he gave a paper in Paris reiterating these claims, and was immediately attacked for misinterpreting his results. Critics suggested that the cultures were almost entirely of connective tissue cells and the thyroid cells were contaminants. Carrel, in his biographers' words, 'tried to overwhelm his critics by showmanship'. He set about cultivating chick heart cells in culture for as many generations as possible, and produced an 'immortal' heart strain which became famous worldwide. This was first

reported in April 1912 when it was eighty-five days old, and featured in countless subsequent accounts of Carrel's work. The culture survived many passages, eventually outliving Carrel himself, and by the end of 1912 was well
established as a heart cultivated *in vitro*, and a rebuttal of the sceptics'
arguments.[33]

Carrel's fame reached new heights in 1912 when he was awarded the Nobel
Prize for physiology and medicine. This was an even greater stimulus to visibility than it is today, and generated an immense amount of publicity among
Americans, especially, who were conscious that their intellectual performance
ill-befitted a major power. He was only the third US resident to win a Nobel
Prize since they had first been awarded, in 1901, while Germany, for example,
gathered sixteen Nobels over the same period. After this, Carrel became a
national hero in both France and the US. The citation for the Nobel award
was for blood vessel surgery and organ graft experiments, but it also ensured
further publicity for the tissue culture work. One result was confusion of the
two types of experiment. The discovery that chick heart cells in culture beat
in synchrony led easily to the assumption that a recognisable whole heart was
preserved. Carrel was also investigating the preservation of whole organs by
freezing, with a view to later transplantation.

One product of all three lines of work was a series of reports that hearts
and other organs were alive in the laboratory. The culmination of this confusion was an account Carrel gave to a meeting of the New York Physicians'
Association on 25 October 1912, two weeks after the Nobel announcement.
The *New York Times* conjured up a scene redolent of a Bernardian laboratory
brought up to date, reporting that after

> many weary months of progressive experimentation, [he] had before him
> in his laboratory a living 'visceral being', living though totally severed . . .
> from the brain . . . There, under the very eyes of the eager wonder-worker,
> was a dog's heart beating its 120 beats a minute, just as though nothing
> had happened, a dog's stomach digesting food . . . a dog's intestines and
> kidneys functionating [*sic*] as though the surgeon's knife had never been
> near . . . an entire system of organs alive outside the body.[34]

This story was repeated, in more or less garbled versions, throughout the
American press. Generally speaking, the papers more remote from the major
metropolitan centres carried the wilder exaggerations. For example, the
St Paul, Minnesota *Rural Weekly* in an article which also spoke of head

transplants, stated that 'it is said that he [Carrel] could by taking various parts of different animals, build up a new living creature'.

These experiments fitted in with the picture that was emerging before the public of the growing powers being attained by the 'wonder-workers' of biology. Carrel himself emphasised the possibilities for the understanding of control of growth which his work opened up, and made an analogy between the influence of the culture medium on the growth of his cells and the effect of the surrounding fluid on Loeb's sea-urchin eggs. The *New York Times* quoted the Frenchman's own words from the *Journal of Experimental Medicine*:

> The growth of the tissues of mammals is probably controlled by the conditions of the interstitial lymph in the same way that the growth of the sea-urchin is influenced by the condition of the water . . . It may be assumed that the facts discovered by Loeb in the lower marine organisms are the expression of general laws which control the development of the tissues and organs of the higher animals as well.[35]

The approving reference to Loeb's work is surprising in view of Carrel's later public dismissals of 'the illusions of the mechanists of the nineteenth century, the dogmas of Jacques Loeb, the childish physico-chemical conceptions of the human being'.[36] Carrel was radically out of sympathy with the mechanistic temper of the Institute, and worked in isolation from most of his colleagues.[37] Yet this was not the only time he found it expedient to ally himself publicly with the mechanists' programme. Carrel joined Loeb on a Rockefeller-funded visit to France in 1912, and found himself interviewed following Loeb's assertion – newly sensational to European ears – that only 'technical difficulties' hindered the artificial preparation of living beings. He appeared to offer unqualified support, saying, according to the *New York Times*, that

> The day when the creation of protoplasm shall be achieved is not far distant. In order to study more minutely the mysterious mechanism of life we have been busy in our laboratories and using new instruments of research . . . The day will come, very probably, when science will force into the light of day the mystery of life. On that day we shall be able to create living beings.[38]

On the face of it, this can only be explained as either a misquotation or a product of Carrel's loyalty to the Institute and his expatriate's desire to discomfit his conservative compatriots. Either way, it was a very public linking of the two men's work and opinions. As the *New York Times* commented, Loeb and Carrel's worldwide reputations gave their utterances great weight. The paper quoted a French listener so enthused that he foresaw that 'we shall be able to fabricate blades of grass and even animals'.

The link, once made, was widely reproduced. A rich example is a long article which appeared in the British news magazine *World's Work* in 1913. The article, which centres on Carrel, has a punning title, 'On the trail of immortality', referring both to the Nobel award and the tissue culture experiments.[39] The text describes all the things Carrel had done, or was reputed to have done – surgery, tissue culture, cancer research, work on relations between cell death and the death of the organism. The 'visceral organism' experiment is described, as is the putative connection with Loeb's work, in almost the same words as were quoted in the *New York Times*. Carrel was positioned

upon the side of those who believe that life is purely a matter of chemistry; that there is no such thing as 'vitalism', but that all living phenomena have a natural and ascertainable explanation. He has not created life, it is true; but, by purely chemical processes, he has been able to rejuvenate a dying cell and keep it indefinitely alive. He has not demonstrated that life itself can be created chemically, but he has practically shown that immortality is merely a question of chemical reaction.

Four portraits illustrate this article – of Loeb, Ross Harrison, Carrel and Edward Shafer. And the work and public utterances of all four are taken as reinforcing the credibility of the claims made. In articles like this, Carrel's work was incorporated into a myth of a collective scientific endeavour to create life, a myth which symbolised the power of the new experimental biology to control living phenomena. How did lay commentators respond?

Public responses to the 'creation of life'

Newspaper science reports in this period were sparse, so any attempt to infer public response is based on a rather small number of examples. However, it is possible to offer some tentative generalisations for the span from the turn of the century to the eve of World War I. Moreover, the fact that there were few professional science writers means that what comments did appear were often written by regular editorial writers, who may have been more concerned with their readers' opinions than full-time science writers of a later date.

When such commentators turned to the consequences of the new developments, they most often saw them in the same positive light as the rest of science. This was a time when, as the *Nation* saw it,

> We believe that there never was so general an impulse to pay honor to science and to do homage to the scientist . . . It is part of our proud consciousness of superiority over the past. The advances which civilization has made are perceived to be largely the work of science, pure and applied.[40]

Of the two, applied science drew higher praise, and the practical benefits of the new biological discoveries were often endorsed in the press, frequently with quotes from scientists. When Delage announced his advances on Loeb's work with frogs, the *New York Times* reported that

> Physiological processes are being chemically imitated with so profound effect upon living cells that Prof. Richards of Harvard, than whom there are few more conservative chemists, recently exclaimed,
> 'When one realises that our frail and often jangling human mechanism is actuated by a series of chemical reactions . . . one feels that chemistry must still have vast treasures in store for the human race. What may she not accomplish for the comfort of the living, for a rational practice of medicine, for a profound philosophy of life!'[41]

The press were also unstinting in their praise of Carrel's achievements, and optimism about the medical benefits that would accrue from his work marked the flood of reports which followed the Nobel announcement. In Carrel's case, though, many found it difficult to be precise about the benefits, and doctors were either reticent or dubious. 'Those who discussed the matter did so only after stipulating that their names should not be used. Some of these men frankly declared that they did not know what use could be made of the interesting results of Dr Carrel's experiments,' according to one report.[42] But this did not detract from the widespread view that his work was important. After the Nobel announcement the climax of the congratulations came at a meeting in the College of the City of New York at which Carrel was fêted by President Taft, whose splendidly vague claim that Carrel had 'made headway into the secrets of nature and thereby advanced the progress of mankind' was widely reported.

However, there were also comments quite out of keeping with the general, positive attitude to science of the early 1900s. These were mainly of two kinds, not always readily distinguishable. First there were expressions of antimechanism, grounded in a spiritual or religious disquiet about the direction biological science was taking. This kind of protest against the world-view projected by the mechanist experimentalists clearly had roots far back in the nineteenth century or even earlier, but was given greater force by the apparent conclusiveness of the experimental demonstrations of mechanism now being offered. Secondly, there were comments in which early signs of misgivings about how the results of such research might be used are evident. When comment moved from the abstract to the concrete, the growing powers of biology gave pause. Sometimes the two strands were intertwined. The new powers being acquired through biological research would eventually raise new moral problems. At the same time, biological discovery appeared to be helping undermine the traditional basis of morality, which might have offered solutions to such problems. The first outlines of a characteristic twentieth-century dilemma began to emerge.

The growing unease evoked in some quarters by the physico-chemical approach to biology produced some strong language. An editorial in the New York intellectual religious weekly the *Independent* in 1902 spoke of 'a certain feeling of repulsion with regard to the attitude of present day science toward the existence of special principles of vital activity'. The disturbed believer was encouraged to regard this as a passing fad: 'the history of science is a constant repetition of the story of this influx of current scientific thought on the theories which are supposed to furnish the solution of the mysteries of life'. The prevalent model of the living system, having changed many times before, would doubtless change again.

It is not surprising . . . that . . . when the new physical chemistry is beginning to occupy the centre of the stage in scientific thought, there should be a series of fresh biological theories explaining the mystery of life by the latest science.

However, the truth will out:

For the many who feel that life is something superadded to force and matter . . . it will be some consolation to know that their favourite doctrine is only undergoing the usual distortion as reflected in current scientific thought, but is no more likely to be seriously affected than by previous manifestations of this tendency.[43]

Ten years later, the threat to religious ideas seemed greater. A long article in the fashionable pictorial news magazine *Cosmopolitan* in 1912, on 'Creating life in the laboratory', opened assertively:

> Life is a chemical reaction; death is the cessation of that reaction; living matter, from the microscopic yeast spore to humanity itself, is merely the result of certain accidental groupings of otherwise inert matter, and *life can actually be created by repeating in the laboratory nature's own methods and processes!*[44]

The impact of all this is immediately emphasised:

> Think for a moment what this declaration signifies. If it be true, where is theology? If you and I are merely physico-chemical compounds, slightly more complex than a potato, a little less durable than a boulder, what is the basis of our moral code? If man can lump together sand and salt and by pouring water on them create life, what becomes of the soul?

In contrast to this kind of comment, speculations on the possible material effects of the new work were relatively rare. Not surprisingly, they related mainly to Carrel, the surgeon. He attracted plenty of negative comment, for his methods as much as his ambitions, becoming a prime target for antivivisectionists, especially after his Nobel award.

In fact, despite the Rockefeller Institute's best efforts to massage its public image,[45] Carrel appears to have behaved just like the antivivisectionists' stereotype of the unfeeling experimental physiologist. Raised in the tradition of Bernard, he was widely reported to be quite insensitive to his subjects. A Texas newspaper, whch found in favour of vivisection as a 'very ugly but highly useful hand maid of science' nevertheless criticised a speech he made in which he described his experiments on organ transplantation 'good humouredly and with the eagerness of a small boy who describes to papa just what the clock did when it was taken apart'.[46]

Carrel's techniques and his programme both invited speculation about work on humans. This in turn gave rise to suspicions about human experimentation, suspicions with which the Institute was familiar: 'a hospital beside a research laboratory known to conduct experiments on animals was *ipso facto* suspected of experimenting on human beings'.[47] Carrel himself often made known his hopes of extending his work to humans. Other scientists suggested how this might be done. The *New York Sun* reported in December 1912, shortly after Carrel's Nobel award was announced, that the director of the Pasteur Institute had proposed taking the spirit of Britain's Anatomy Act a stage further, believing that

> the great need of the day in experimental medicine and surgery is to have a supply of live human beings to experiment on. I have no doubt that with plenty of human material . . . wizards like Carrel and Flexner would within the next two years solve almost all the problems which are yet baffling the medical profession.

He suggested that the bodies of condemned criminals should be made available for this purpose, as an alternative to wasteful execution and burial.[48]

Carrel's own transplant experiments were often linked with humans, even if only at the level of simple denial exemplified in the *New York World*'s report on an experiment in which, supposedly, Carrel 'added to the wonders of his surgery the transfer of a heart, stomach and kidneys from their casement of flesh to the vessels in his laboratory'. The writer added: 'none of the organs, of course, is that of a human being'.[49] Other newspapers were less reassuring. The *Buffalo News* related how Carrel had 'not only transplanted organs of animals so that they perform the same function in human beings, but has cut out organs and grown them to monstrous size in hermetically sealed jars'.[50] The *New York Press* found another twist to the story:

> Recent discoveries in science tend to refute the Darwinian theory rather than confirm it. Various sections of the anatomy of the monkey when transformed [*sic*] to the human body in surgical operations, do not thrive as well as those organs taken from some of the lower animals, such as the sheep, the dog and the cow.[51]

This information is attributed to Carrel, and the copy of the article from the Rockefeller archive bears the comment 'false interview'. Similar fabrications multiplied. A week later the *Buffalo Express* told how 'in a refrigerator of the Rockefeller Hospital . . . may be seen sections of living bones, pieces of living muscle, bits of liver, of veins, of arteries, of kidneys, of almost any part of the human body'.[52] This Frankensteinian vision was in an article praising Carrel's achievements. But derivatives of the original *Frankenstein* tale had already progressed from the reanimation of cadavers to the use of artificial organs. As early as 1900 a story appeared of a scientist working with artificially produced human organs and joining them together by a network of vessels. This kind of image now seemed to be turning into reality.[53]

'Life as per recipe' – the British Association meeting of 1912

The press responses examined so far are mostly American, but on one occasion these ideas attracted a large volume of instant comment in Britain. The responses on this occasion – the presidential address by Professor E.A. Schafer to the British Association for the Advancement of Science meeting at Dundee in 1912 – are instructive both because a natural scepticism in the face of claims that life could be created in the laboratory seems to have been more widely overcome in this case, and because extrapolations to the level of control over human life were more frequent.

The work of Loeb and Carrel had already been described to British newspaper readers from time to time, but the BA meeting, in keeping with long tradition, was the occasion for a much wider discussion.[54] The space given over to science during the meeting, and hence its public impact, was considerably greater than for comparable meetings in the US at the time. Newspapers there did not show as much interest in the annual gatherings of the American Association for the Advancement of Science until after 1920.[55] It may thus be easier to assess opinion in Britain from this single episode than from a longer series of reports on different discoveries in the United States.

Schafer was one of the three leading figures in the establishment of experimental physiology in Britain, along with Michael Foster and J.S. Burden-Sanderson, all of whom studied at University College London under William Sharpey. Schafer rose to the chair at University College and in 1899 was elected to the chair of physiology at Edinburgh, where he stayed until 1933. His BA address was a brief episode in an influential career.[56] He used it, as was expected, to expand on the future of his discipline in a forum where it was legitimate for a senior figure to be a little speculative.

In 1912, this scientific event of the year had spectacular advance publicity:

> Professor Schafer had let it be known that his address would be concerned with the chemistry of living processes, the gradual passage of chemical combinations into the condition which we call 'living', and the possibility of bringing about this passage in the chemical laboratory.[57]

In his address, heard by over two thousand people, his main argument was that biological discovery was tending to blur the distinction between living and non-living matter, so that

> The problems of life are essentially the problems of matter . . . the phenomena of life are investigated and can only be investigated, by the same methods as all other phenomena of matter, and the general results of such investigations tend to show that living things are governed by laws identical with those which govern inanimate matter.

He drew on a wide range of work to support this claim, including especially Loeb and Carrel's experiments.[58] The description of recent work on the endocrine system also impressed many in the audience. The *Daily Mirror* correspondent, under the heading 'Scientists and the secret of life', wrote that

> One of the professor's most fascinating statements was that 'hormones', although responsible for some of the most vital characteristics of life . . . have been found, when analysed, to be of an extremely simple chemical character.
> One of these 'hormones' – which one might almost describe as one of the secrets of life – has been actually prepared from chemical substances in a laboratory![59]

After this, Schafer commented on the possibility of synthesising some of the components of the cell, in a way which illustrates the simplified conception of cellular chemistry some physiologists held at this time. For example,

> The researches of Miescher . . . have shown that . . . a body so important for . . . the cell as the nucleus – which may be said indeed to represent the quintessence of cell life – possesses a chemical constitution of no very great complexity, so that we may even hope to see the material that composes it prepared synthetically.[60]

After a similar comment on proteins, he came to the point:

> The elements composing the living substance are few in number . . . The combination of these elements into a colloidal compound represents the

chemical basis of life, and when the chemist succeeds in building up this compound it will without doubt be found to exhibit the phenomena which we are in the habit of associating with the term 'life'.

Not pithy, but warrant for the universal interpretation that he was asserting that life could be created in the laboratory.

And what was the reaction? Schafer's remarks undoubtedly commanded public attention. He carefully collected all the press cuttings he could find, and the archived scrapbooks he kept of the meeting amount to some 440 pages.[61] Nearly all the British papers were well satisfied that the speech was sensational; the exception was the *Manchester Guardian*, whose correspondent suggested that it contained nothing new. Even that paper, like all the others, printed long sections of the address. In every case, Schafer's remarks were taken as authoritative, and the idea that life might be created was accepted. Other scientists endorsed his position. Sir Ray Lankester, one of the best-known zoologists of the time and a protégé of Huxley who emulated his mentor as a populariser, was quoted as saying that those who were engaged in research were in complete agreement with Schafer. In his regular column in the *Daily Telegraph*, Lankester himself wrote that Schafer 'gave an admirable and clear statement of the progress during recent years towards . . . the con-struction in the laboratory by chemical methods of the complex chemical combination which we call "living" '.[62]

From the other side of the mechanist/vitalist divide, there was unexpected support for Schafer from Sir Oliver Lodge, the physicist, radio pioneer and president of the Society for Psychical Research. Lodge was very well known for his popular articles and lecture demonstrations, and was deeply committed to both physics and spiritualism.[63] He did not believe that an understanding of the material basis of life could undermine either vitalism or his own reli-gious beliefs. Indeed, in an interview quoted in at least four papers, he reminded his questioner that 'he himself had suggested that possibly living material would be artificially obtained in the laboratory one day'.[64]

Most editorial comments during the following week pondered the con-sequences rather than the plausibility of life creation, and there was a wealth of negative comment. The more believable these writers found the idea, the less they seemed to like it. Some took it that this life would be biologically basic, some type of protoplasm or 'colloidal slime'. The *Daily Mail* compared the making of synthetic protoplasm to synthetic rubber. Readers were assured that it is a long way from synthetic rubber to an indiarubber, and it is still further from a speck of artificial protoplasm . . . to a man or a mollusc or even a microbe. We may eventually be able to produce by chemical and mechanical means a living substance, but that is a very different thing from creating a living organism.[65]

This meant that any fears raised by the prospect of creating life were premature. The *Spectator* was quite explicit on this point, suggesting that 'it is some consolation to think that the life-producing scientist, whose advent Professor Schafer confidently anticipates, will be no Frankenstein. The utmost he can produce is something many millions of years removed from the level of humanity.'[66] Nevertheless, there was a general distaste for the idea. The *Daily Telegraph* implied this by observing that the address was 'pervaded by an air of unconcern . . . that will produce a curious sensation in the mind of an ordinary reader'.[67] The same paper assured readers that, if successful, Professor Schafer's scrap of living matter 'would not be a Frankenstein by a good many millions of centuries'.

Others invoked Frankenstein as a warning, rather than a kind of reassurance. As with so much biological research later in the century, Schafer's suggestion was weighed up according to whether it did, or did not, resemble the projects of fictional scientists. The *Yorkshire Post*, for example, opined that 'The upshot of all this is that serious sanction is given to the wild imaginings of the romancers who have pictured the man of science – usually represented as a villain – manufacturing a living organism, which commonly turns out a Frankenstein monster.'[68] The *Manchester Guardian*'s correspondents comment that if 'we should be able to turn out living germs of our own creation – wholesale from the test-tube in which they had been made', this would be 'like the homunculus in Part II of Goethe's *Faust*, nay, if they proved noxious, it would be a new version of *Frankenstein*'.[69]

The *Leeds Mercury*, on the other hand, asserted on the same day that 'there is no danger of a new Frankenstein monster', as if the old one had actually lived at some time in the past. Even the *Lancet* discussed Schafer's ideas at length in these terms, contrasting his notions with the dreams of the alchemists, with Paracelsus' recipe for a homunculus, and with 'monstrous beings after the sort fashioned by Frankenstein'. There was not even the prospect of 'synthetic beefsteak', the journal said: 'The most that we can look for is no more than to create the dim beginnings of vital activity, to mould the faint transitional forms between living and non-living matter.'[70]

But despite such reasssurances, the popular assumption was that creating life would eventually mean creating people. Perhaps they would only be miniature laboratory people, as light-heartedly depicted by the cartoonist for *Modern Man*. But the prospect was in any case disquieting. *The Times* suggested that the idea of artificial creation 'is publicly discussed by men of science, most of whom, no doubt, hope that the professor is right; but it is also privately discussed by the general public, most of whom hope that he is wrong'.[71] And if, as this implied, the feelings of scientists and public were directly opposed, this was because the public supposed that 'if it became possible for human beings to produce living material, the last mystery would be gone from the universe'. The London *Evening News* saw a similar distance between public reaction and scientific determination to pursue such work. The paper declared that most people found the notion of man-created life distasteful, blasphemous, and suggestive of meddling with forbidden things: 'however . . . no such scruples are likely to deter those who have entered on this branch of scientific investigation'.[72]

The question of consciousness or, as it was more often put, the soul, was much discussed when creation of higher organisms was considered. The *Daily Mirror* took up this theme at some length. The leader-writer suggested that the strength of the claim that life could be created derived from the new-found authority of science. 'To the chemist the majority now more confidently look for truth than they once did to the priest. What the scientist says is so, because he says it.' On the assumption that life creation would be achieved, the writer imagined how 'In time, this *thing* might, by industrious coaxing, be induced to become an *animal*, according to the accepted manner of "transformism", and evolve new capabilities and wants' (original emphasis). The ultimate biological achievement was, in fact, now conceivable.

> Then, at last, after much more coaxing, a person, a rudimentary savage person might emerge; and so the whole process of evolution be conveniently shortened, in lecture time, by an ingenious gentleman on a platform. 'Ladies and gentlemen, I propose to turn this mass of weltering matter into a man.'

But what kind of life would this be?

> More life the scientist reveals to us; does he give us more life, the only life that matters to us, our life, the life of man, and of the soul in him?

The rhetorical question is answered:

> He [the scientist] patches us up and we thank him; he wards pain from our bodies and we are pleased; he fits up our life with many useful devices, for which occasionally we make him rich. And now – perhaps – he will make more life for us out of the elements. Only let us be clear about that ambiguous word. The 'life' he will make will not be of any importance to us, because it will not be a deepening of quality in the soul.[73]

One reader, at least, shared this view, writing that 'If the rapid advance of modern science ever does bring things to such a pass, that life – human or animal – can be created at man's will, the results emanating from such a creation would . . . be too awful to contemplate.'[74]

These religious or quasi-religious misgivings evoked other, more strongly worded, expressions of antagonism toward science in general. An article in the weekly illustrated paper the *Graphic* complained bitterly of the learned professors 'who blink out upon the world with eyes that have been tired with the microscope and have done their best to rob me of any little self-respect which I bore up against the adverse influences of the weather and the world's minor woes'.[75] The piece gave a sketch of the intrusive methods of the new science as it impinged on the integrity of the body:

> They have plucked the flesh off my bones and thoroughly examined it with strong glasses. They have put their fingers upon our quivering nerves and made a chart of our sufferings and sensations. They have peered into the living cells of our mortal flesh, into the folds of our throbbing brains. They have analysed our chemical constituents, they have searched out our souls with their scalpels. And the result? . . . If they are to be believed (forsooth) the result is to trace back the ancestry of man to the lifeless chaos of nature's first stink pots.

Rhetoric like this is part of a clear line of descent from William Blake to Theodore Roszak. A scarcely suppressed hostility and fear pervade the entire article. For worse is to come:

> They are clever, these chemical compounds who call themselves professors. They have made a pulp of the human brain and body and dealt with it as apothecaries, with measuring glasses and crucibles and little phials, and

nicely balanced scales. They know so much about it now that Professor Schafer . . . is able to make out a neat recipe for the preparation of life's quintessence . . . he sees no reason why in a little while the chemist should not put together these constituents and create the nucleus of cell life.

There is no trace here of the warm feelings toward scientific progress evident elsewhere. The proper response to scientists' aspirations is ridicule:

All the great worlds of mystery that lie outside the laboratory . . . have been ignored by these apothecary philosophers . . . must all our poetry, all our spiritual yearnings, all our wide and limitless realm of the imagination, all our faith in things immortal, be interpreted by chemical actions and reactions? The world's laughter rings out above the prattle of the apothecaries. A sense of humour saves us from despair.

This example shows very clearly the strength of feeling evoked by the spiritual and moral implications some saw in the artificial production of living matter. Schafer's address also prompted comment on the possible material consequences of such a feat, and again a negative reaction is discernible.

The most impressive example in this area is a remarkable editorial which appeared in the *Daily Herald* on 6 September 1912. The *Herald*, founded the previous year, was trying to become a popular paper of the left, and the bulk of the week's issues were devoted to the Trades Union Congress meeting which overlapped with the British Association. They nevertheless found space for a comment on Schafer.

The article, headed 'Life as per recipe', begins with a brief description of the way in which, according to Schafer, life may be created. This is accepted as scientific fact, and is immediately extended to the highest level. 'This idea of artificially creating life . . . may lead to the manufacture of living men.'[76] The writer is unequivocal about the consequences. The living men will be created

in a nice, clean laboratory, under Government supervision and control. (This sort of thing is always under supervision and control.) Nice, well-intentioned scientists will successfully perform, in fact, what Frankenstein achieved only in fiction. Can't we hear them talking? 'Better give this one a little more bicep: looks as if coal-miners are running short'. Or 'You're allowing that man too much brain space; he'll be joining a trade union if you don't take care'. Or 'Are you sure he won't be able to have children? It'll be confoundedly inconvenient if he does.'

This strikingly well-developed vision of control, prefiguring Aldous Huxley's imagery of twenty years later, is used to bring the whole application of science into question:

> These imaginary scraps of conversation are not so very imaginary; for what purpose is the new discovery most likely to be utilised if it can be carried out to its logical conclusion? To the making of men, to the making of national backbone? No. The new discovery will be exploited, as every new discovery of modern science has been exploited, immediately contrary to the best interests of the nation. Elderly duchesses will be kept alive, their pug dogs will have their dying hours delayed, and the greatness of science poured out upon the least worthy.
>
> As to the workers . . . it is easy to see what may happen. The life creators will join with the eugenists. A section, and probably a large section of the working classes will be forbidden to have children . . . The necessary population, calculated in Whitehall, will be supplied with all their virtues and vices – or shall we say their virtues, and leave it at that? – ready made. The slave foredoomed will always be a slave; the scientists will have seen to that.

The reaction to eugenics is one which still echoes in debates about biological control. But there is also a more general current of criticism worth noting here. The piece concludes with a question which would be heard more often as the century grew older:

> Why does not the British Association . . . provide some remedy for the evils of the present, instead of adding to the difficulties of the future? That is the question which every intelligent working man must be asking himself on reading the sensational tidings of the latest developments of modern science.

Conclusion: creating life and controlling people

How representative is this episode of public debate about biology in the twentieth century? I have concentrated here on press comments on the work, or representations of the work, of just a few researchers. However, the remarkable fact that the claim that scientists were on the verge of creating life in the laboratory was taken seriously in the early years of this century does give access to a perhaps unexpectedly wide range of direct, public responses to experimental biology. These responses demonstrate a number of things.

First, just as the idea of biotechnology has a long history which is significant for our understanding of the term today, as Robert Bud has described, so does the public imagery associated with what would later come to represent the 'biological revolution'.[77] This was not just a time of fictional reworkings of what scientists might be said to be up to, as in Mary Shelley's reference to rumours about Erasmus Darwin. Here were serious scientists making large claims of a kind which would not regain such plausibility until well after World War II.

Secondly, the style of debate, and the way it is represented in the print media, is not far distant from the treatment of such issues today. A mass reading public, a diversified and stratified press, and, as I have suggested, researchers who were at least to some extent prepared to behave as 'visible scientists', were all in place by the years between 1900 and 1910, in Britain and the United States.

Finally, and I suggest most significantly, these responses show the power of developments in biological science to evoke profound disquiet at a time when publicly expressed attitudes to science and technology tended to cover a narrow range – from millennial celebration to cautiously optimistic progressivism. The power of the new experimental biology to tap feelings of deep ambivalence is now very familiar. The fact that it was apparent at such an early stage in the development of modern biology helps support my suggestion that present-day responses are not the product of the rise of any generalised 'anti-science' sentiment. They are more likely to be evoked by specific reservations about biological manipulation and its potential applications, just as they were when Loeb, Carrel and Schafer were helping to shape the popular image of life in the laboratory.

Chapter Five

Into the Brave New World

Anyone who takes a story of science which starts in some earlier period into the twentieth century faces a nice problem of strategy and selection. Among the many transformations of science and technology in our century, speed and scale are most apparent. There is simply more science, more scientists, and more discussion of their work. Our awareness of science, and of the infiltration of modern technology into every corner of everyday life, is greater. And there is far more discussion of science and technology, often in new media with new audiences.

So while selection is all in any historical writing, from here I must be still more selective. In this and the next two chapters I consider the outlines of discussion of biology in the first three-quarters of the twentieth century, trying to register the continuities as well as the novelties. I will focus on a small number of key texts, but place them in the context of a larger, if highly compressed, outline of the general line of development of public debate about the life sciences.

In this chapter I consider the inter-war period, a time when the high hopes for biology in the pre-war years – the expectation that life would be created in the laboratory – were damped down by a realisation that life harboured hitherto unknown complexity. As an editorial in the *New York Times* put it in 1936: 'The deeper the chemists probe into the cell the more baffling it becomes. The fertilisation of the seed, the growth of the embryo, the development of an individual, the transmission of features to the next generation, adaptation to the environment, evolution of lower into higher forms – it is this that is life, this that defies the mechanists and vitalists alike.'[1]

At the same time, though, there was a ripening perception that experimental biology was the science with the greatest potential to transform human life, a perception which dominates the speculative literature of the time, and which we still express in the terms laid down in the 1920s by J.B.S. Haldane and Aldous and Julian Huxley. The milieu in which that perception arose was a rather different one from that which gave rise to Loeb, Carrel and Schafer's speculations about the potential powers of biology. Turn of the century optimism about science and the human condition had been called into question by war and depression. The possibilities of science, and its potential social effects, were more widely debated. How did biology figure in that debate?

The shape of post-war science

The end of World War I, the start of Hobsbawm's 'short twentieth century',[2] was a turning point for science as for so much else in modern life. The 'chemists' war' convinced governments of the strategic importance of research, paving the way for greater state support. The deployment of poison gas, combat aircraft, tanks and submarines was a grave blow to pre-war techno-logical optimism. I will return to this shift in attitudes to science and tech-nology after a sketch of the institutional and scientific background to public comment on biology.

In Britain, changes in medical research policy were put in hand just before the war. The state-financed Medical Research Committee was set up under the 1911 National Insurance Act. After the war, the committee was hived off from the new Ministry of Health and reconstituted as the Medical Research Council. It was to perform for medical research the role of the newly formed Department of Scientific and Industrial Research (DSIR) in industrial research. The MRC was thus the first of the British research councils to assume its modern shape.[3]

The new council was committed from the start to basic research, much like the Rockefeller Institute in the USA. Its concept of the goals of an organ-ised system of biomedical research was shaped in its early years by men con-vinced of the need to pursue a fundamental understanding of living matter through physico-chemical experimentation. Michael Foster, who had worked up a broad programme for general physiology which embraced histology, embryology and 'chemical physiology' or biochemistry, brought Frederick Gowland Hopkins to Cambridge in 1898 to help forge biochemistry into an academic discipline. When Hopkins was consulted about the secretary's position on the original Medical Research Committee he proposed the similarly inclined Walter Morley Fletcher. Fletcher was duly appointed and served for the next twenty years as the main orchestrator of biomedical research in Britain, and as a convinced promoter of biochemical approaches.[4] His influence extended to private as well as public funds. In 1924, for example, he secured a bequest of £200,000 for Hopkins's new biochemistry laboratory in Cambridge.

In the US, biomedical research still relied mainly on private dollars, but the Rockefeller and others sustained a flourishing research community. On both sides of the Atlantic, there was a vigorous development of the pre-war achievements of Mendelian genetics and, later, the outlines of classical biochemistry. These went along with changes both in the attitudes of leading scientists to their work and in the public image of their science. Experimentalism was now firmly established, but there were few scientists who espoused the extreme mechanistic materialism of Loeb and his followers. Garland Allen suggests that there was a shift in philosophical allegiances in physiology toward what he calls 'holistic materialism', exemplified in the work of Sherrington on nervous integration and, later, of Henderson and Cannon on physiological equilibria.[5] This shift is less discernible in genetics, dominated by Morgan and his colleagues, or in biochemistry, but it is certainly true that sweeping pronouncements about the ultimate nature of living things are harder to find. Paul De Kruif's admiring *Harper's Magazine* profile of Loeb in 1923 reported the latter's regret that he had failed to convert his colleagues: 'He bewails his isolation. He complains of the fact that he grows old with the encouraging support of so few of his scientific contemporaries.'[6]

The kind of success researchers were having made it seem less likely that a complete understanding of life was at hand. A mass of new information fostered a new appreciation of the complexity within the cell. Each fresh success disclosed a new set of experimental and theoretical problems. Every new discovery of a vitamin, hormone or gene made the organism seem more complicated, more delicately balanced, more difficult to describe as the product of any single organising principle. Advances in these areas also pointed up some unsolved problems – with hindsight, the absence of a satisfactory chemical account of heredity and of a coherent picture of the structure and properties of macromolecules seem especially important.

So a more general awareness of the complexity of living things encouraged a more measured appraisal of the prospects for biological advance than had featured in public discussion before the war. The sense of a new appreciation of the intricacy and refinement of the machinery of living things was well expressed in *The Science of Life*, a three-volume review for the lay reader by H.G. Wells, his son G.P. Wells and Julian Huxley, first published in 1930. After a lengthy description of the genetics of *Drosophila*, they suggested that 'the discoveries we have been narrating reveal that *the work of making a fly is ... more astonishingly delicate and complicated than we could ever have imagined.* If a fly an eighth of an inch long needs several thousand separate and different units to conspire in building up its structure, how many will prove necessary for a man?'[7]

Much the same feeling is apparent in an earlier comment by E.B. Wilson, in a well-known address of 1923 in which he took up T.H. Huxley's discussion of the physical basis of life. He combined an enduring, though undogmatic, commitment to mechanism with the realisation that new problems would take time to unravel. He was candid about biologists' ignorance about the nature of biological organisation and embryological development, although this did not mean that his 'faith in mechanist methods and conceptions is shaken'. As he argued:

It is by following precisely these methods and conceptions that observation and experiment are every day enlarging our knowledge of colloidal systems, lifeless and living. Who will set a limit to their future progress? But I am not speaking of tomorrow but of today, and the mechanist should not deceive himself in regard to the magnitude of the task that still lies before him.[8]

This confidence tempered with realism meant that a goal like the creation of life – which stood as an implicit test of biological understanding – was not repudiated but was now seen as an aim for the distant future, not an imminent achievement. Again *The Science of Life* offers a representative post-war view. After a materialist account of biogenesis, the two Wells and Huxley endorse the idea that living matter might one day be created artificially. But they emphasise that this will not be easy:

To be impatient with the biochemists because they are not producing artificial microbes is to reveal no small ignorance of the problems involved. Living matter is matter; but it is quite appallingly complicated matter, many times more complicated in its construction than any other substance known anywhere in the universe ... We rightly praise the skill of the chemists who build up dyes and drugs to order, but to build up living matter, substances as complicated as their highest achievements in synthesis would have to be used as the basic bricks. In any attempt to create living matter, we begin about where the modern chemist leaves off.[9]

Although the scientific picture had changed, there were still sporadic press reports in the established vein of life-creation stories. But the absence of authoritative support meant that they tended to fade away relatively swiftly. A good example of this change in climate is the response to F.G. Donnan's talk at the British Association meeting of 1928. Donnan, a physical chemist, described the work of the biophysicist A.V. Hill on respiration. In the talk, Donnan's interpretation of this work, mainly concerned with mathematical description of oxygenation–deoxygenation equilibria in the blood, was that his colleague was 'on the eve of discoveries of outstanding importance'. Speaking to reporters later, he went further, suggesting that 'the continuous analysis of the phenomena of the living cell . . . must lead to such an understanding of the organisation of life that there is no reason why the creation in the laboratory of a living cell of the simplest kind, on the physical plane, should not be effected'.[10]

Plainly, this was a 'prophecy that scientists will create life', as the *Daily Mail* proclaimed the next morning. The widespread accounts of Donnan's speech, and the press's subsequent pursuit of Hill on holiday in Devon, show that the idea was still news. But the later course of the story was very different from the response to Schafer's similar suggestion in 1912. Hill's denials were unequivocal. He quickly put his view that 'it will be a very long time before we shall begin to understand the general lines on which the living cell was made up, and almost hopeless to attempt to put them together'.[11] The next day, he reiterated that it was 'nonsense' to say he was on the verge of any great discovery.[12] This rapid denial of Donnan's forecast was reinforced by other scientists asked for comment, who either refused to say anything or waxed extremely sceptical.[13] Editorial comments underscored this scepticism, and the story died.

A similar response greeted the later work of Herrera in Mexico, reported several times in the 1920s and 1930s. Before the war, his experiments, like Butschli's, had been seen as one more instance of the effort to create artificial protoplasm.[14] Now his work, mixing organic oils and solvents with water and observing microscopic movements in soapy droplets, was seen simply as a diverting curiosity. It could be reproduced by any amateur, and no one could now credit that so simple a demonstration could lead to formation of living things. Herrera's bodies were now described as imitations of life, pseudo-cells.[15] Herrera himself eventually decided that he really had created proto-plasm, but he was ignored by most biologists. The account of his results in *Scientific American* in 1931 was full of doubts and qualifications. It was a dying echo of the life-creation stories from before the war, and the work bore no resemblance to contemporary research in physiology or biochemistry.[16]

In general, then, up-to-date research in biology was less often at the centre of attention in the inter-war years than in the first decade of the twentieth century. Sensational stories about breathtaking developments were the exception rather than the rule. More typical were comments like that of the *New York Times* science correspondent in his annual review of scientific progress at the end of 1935, who told his readers that 'in the field of biology there is much groping'.[17] This kind of impression meant that many biologists dwelt in happy obscurity. Indeed, C.H. Waddington, who was active in research through the 1930s and onwards, could write in 1969 that

> for the last half century or so, biologists have been used to a rather quiet life, out of the public eye in their academic laboratories . . . The last impor-tant occasion of public excitement about their activities was connected with Darwinism . . . Even the rediscovery of Mendelism and the rise of genetics in the first quarter of the twentieth century produced little inter-est in the public mind. When people have spoken of 'scientists', it has almost invariably been physicists and chemists that they had in mind.[18]

However, Waddington's generalisation is wrong, though it may be read as a fair recollection of the newspaper reporting of science in the earlier part of his career. If we turn to other cultural forms, we can see that the lines of thought about biology which were beginning to take shape in earlier years continued to develop in this period. Biology may not have been in the jour-nalistic spotlight, but its future was still discussed in fiction and speculative non-fiction. It is these which furnish the most important historical examples of public discourse about the life sciences in the inter-war years.

Biology and industrial production – the kneading trough and *R.U.R.*

Most writers about the future of the life sciences now framed their specula-tions around certain common assumptions. They believed that the potential transforming power of biological discovery was important. And the experience of the war and, soon after, the depression made the equation of scientific advance with social progress seem problematic. Whereas pre-war utopians and prophets often took it for granted that technology was beneficial, this now seemed naive.[19]

The most conspicuous products of this reappraisal were the literary dystopias, but there was also a growing literature in which the pace of progress, the quality of future life, and technology and social change were discussed in more detail. By 1939, Waldemar Kaempffert, the redoubtable science correspondent of the *New York Times*, was telling readers of his book *Science Today and Tomorrow* that he was contributing to 'a whole literature on what is called "the impact of science on society" which has been produced within the last decade'.[20]

Well before this, however, a piece of drama had made its mark on public awareness of science. The Czech Karel Čapek's play *R.U.R.* or *Rossum's Universal Robots* was a worldwide theatrical sensation in the early 1920s, and has the distinction of introducing the term 'robot' into English. *R.U.R.* is a work of pure fiction and, like *Frankenstein*, is partly a comment on the transmutation of a scientist's aspirations in the real world. Clarke exaggerates somewhat when he suggests that it be seen as 'the earliest imaginative adjustment to the 19th-century expectation of constant progress',[21] but it was certainly one of the first to reach a mass audience. Like Mary Shelley, Čapek uses a biologically based image to dramatise the relations between people and their technology, but this time the setting is industrial. The way this works out represents a significant crystallisation of post-war pessimism, a synthesis of several diverse strands of criticism of science and industrialism.

The key novelty in *R.U.R.* is the marriage of the artificially created humans with the production line. This enabled Čapek to develop symbolic possibilities some have seen in *Frankenstein*. As one commentator put it in the 1970s, Frankenstein's monster is 'the perfect proletarian. Harmless, roving, cut off from the bourgeois ties of family and place, he is not merely an exploited victim of the production process; he is *himself* a product.'[22] Now the product was to be mass produced.

Creation of artificial people in *R.U.R.* in fact begins with a Frankenstein figure. Old Rossum, attempting the chemical synthesis of protoplasm, discovers a living substance unknown to nature. His only motivation was to prove that life could have been created without divine intervention, and after his discovery he spends ten years striving to fashion a man. Enter his thrusting young nephew: 'It's absurd to spend ten years making a man. If you can't make him quicker than nature you may as well shut up shop.'[23] The younger man redirects the research, undertaking to 'overhaul anatomy' on sound engineering principles. Thus are the concerns of the nineteenth-century materialist displaced by a purely industrial rationality.

This was Čapek's intended reading, as he explained in the *Saturday Review* in 1923.[24] The new men of simplified design – the robots – are thus turned out on a production line like Ford's Model T, for the profit of the firm. They are sold as the benefactors of mankind who, freed from toil by these willing servants, will finally attain a life of ease and creativity. Their designers are tremendous enthusiasts. Čapek furnishes a graphic description of the factory as the general manager tries to impress a visitor:

> I will show you the kneading trough . . . In each one we mix the ingredients for a thousand robots at one operation. Then there are vats for the preparation of liver, brains and so on. Then you will see the bone factory. After that I'll show you the spinning mill . . . for weaving nerves and veins. Miles and miles of digestive tubes pass through it at a stretch.

The outcome of all this industry is not the utopia the proprietors of the firm have proclaimed, but a world of confusion and despondency in which robots throw men out of work and are used to fight nationalistic wars. Eventually the robots rebel against their masters, and destroy an infertile and demoralised humanity. The ending of the play is ambiguous – a male and female robot, designed for sterility, nevertheless fall in love and, it is implied, start to perpetuate their new artificial species. But the moral is clear. Faustian meddling now brings disaster for the whole society, not just the individual scientist.

The play was first performed in the National Theatre in Prague in January 1921, and within a few years had been translated into a variety of languages and performed throughout Europe and in the US and Japan. According to his biographer, Čapek became a household name and, although *R.U.R.* was his least favourite work, 'it is safe to say that no other play on modern technology has so captured the public's imagination'.[25] The key to its appeal may be the symbolic weight of the robots. Čapek's clear vision of what he was trying to do produced a readily intelligible combination of ancient and modern. Later robots, fictional and real, were machines, the product of electronics and micro-engineering. Yet Čapek's robots are quite clearly organic, what we now call androids. But until 'android' entered the lexicon, 'robot' was used to denote any soulless analogue of a human being, whether biological or mechanical. This ambiguity stemmed from the powerful combination of images in the play, uniting a number of criticisms of science, technology and mechanisation, of scientific aspiration and of the regimentation of a machine-dominated mass society. The development of biology in capitalist industry symbolises the tendency to make machines of men and men of machines.

The robot symbol thus expresses fears of reification, and the problems of relating both to people who treat others as machines, as Taylorism treated factory labour, and to machines that simulate human actions.[26] A more recent locus of these fears has been computer science, particularly artificial intelligence, but this came after World War II. In Čapek's text, the biological association has priority, and the army of thinking machines, androids, clones and cyborgs which populates later science fiction are as much descendants of his robots as they are of Frankenstein's creature.[27]

Biology as perversion – Haldane and *Daedalus*

R.U.R., while symbolically rich, lacked any reference to contemporary research. But there were other works of the period which took up Čapek's themes and were extremely well furnished with details of real biological experiments. These largely came from a remarkable group of British scientists and writers who developed a common interest in exploring the possibilities of biology in the 1920s and 1930s. They included J.B.S. Haldane, initially working in Gowland Hopkins's biochemistry department in Cambridge, his wife Charlotte, a feminist journalist and science populariser, his sister the novelist Naomi Mitchison, and their close friends Julian and Aldous Huxley – grandsons of T.H. Huxley. As literary critic Susan Squier emphasises in a recent study, all five were deeply involved in an intense debate about reproduction which took in sexuality, contraception, motherhood and eugenics. All five, in popular writings of various kinds, offered interpretations of the potential social importance of science in this realm.[28]

Undoubtedly the most influential was Haldane's *Daedalus*, or *Science and the Future*, the first essay in Kegan Paul's series of booklets *Today and Tomorrow*. Haldane's essay shifted attention from rather abstract concerns like creation of life to more concrete possibilities of biological control. His method was ostensibly simple. He extrapolated from work already in progress and suggested a series of goals which he foresaw being attained in the medium term.

The results, however, were extraordinary. *Daedalus* is still remarkable for its variety and scope of suggestion, and as an exercise in disciplined use of the imagination.[29] Haldane's deep knowledge of contemporary biology and his open-mindedness, combined with a frank delight in shocking his peers, helped him paint a vivid picture of possible future developments, each one rooted in fields biologists were already exploring.

At the outset, he stressed that he was making no suggestions rasher than H.G. Wells's proposal in his 'Anticipations' of 1902 that there would be heavier-than-air flying machines, which the reader of the 1920s already knew well had been fully realised. But at the turn of the century, according to Haldane, the centre of scientific interest lay in physics and chemistry. This was no longer true: 'now these are commercial problems and I believe that the centre of scientific interest lies in biology'.[30] He acknowledges that biology is not at a peak of success, but uses this observation to underline its ultimate power: 'We are at present almost completely ignorant of biology, a fact which often escapes the notice of biologists and renders them too presumptuous in their estimates of the present position of their science, and too modest in their claims for its future.'[31]

Earlier forecasters had been too timid:

> With regard to the application of biology to human life, the average prophet appears to content himself with considerable, if rather rudimentary progress in medicine and surgery, some improvements in domestic plants and animals and possibly the introduction of a little eugenics.

Haldane is no ordinary prophet. He launches into a succession of provocative speculations, whose daring is emphasised by his almost gleeful awareness of the likely reaction. He succinctly summarises the potential of a science which manipulates life to disturb social and cultural categories: 'The chemical or physical inventor is always a Prometheus. There is no great invention, from fire to flying, which has not been hailed as an insult to some god. But if every physical and chemical invention is a blasphemy, every biological invention is a perversion.'[32]

What perversions did Haldane foresee? Again stressing that he was only suggesting 'a few obvious developments', he began to outline the history of the future, as seen by a 'rather stupid undergraduate' in 150 years time. First came nitrogen-fixing bacteria, in 1940. This was a great agricultural benefit, after a few teething troubles. When one strain escaped into the sea, 'for two months the surface of the tropical Atlantic set to a jelly'. Eventually, though, this increases growth of plankton and boosts fish harvests. This benign outcome sets the optimistic tone for the rest of the essay. In 1951, he predicted, would come the birth of the first ectogenetic child, conceived and born outside the womb. This had long been a notion discussed in Haldane's circle and, according to his biographer, one he would expand on at every opportunity.[33] Now, he elaborated on the implications, after first establishing that work in this direction was already under way by citing experiments in 1902 in which embryo rabbits were transferred from one female to another. 'Now that the technique is fully developed,' says Haldane's future student, 'we can take an ovary from a woman, and keep it growing in a suitable fluid for as long as twenty years, producing a fresh ovum each month, of which ninety per cent can be fertilised, and the embryos grown successfully for nine months, and then brought out into the air.'[34]

Religious opposition to use of the technique could be overcome by a decline in the natural birth rate in the civilised countries, which would also welcome this much-needed opportunity for genuine eugenic selection of donors. He moves in one leap from simple contraception through the sexual debates of the 1920s to a full development of reproductive technology:

> The effect on human psychology and social life of the separation of sexual love and reproduction which was begun in the nineteenth century and completed . . . in the twentieth is by no means wholly satisfactory . . . However, had it not been for ectogenesis, there can be little doubt that civilisation would have collapsed within a measurable time owing to the greater fertility of the less desirable members of the population.[35]

Later, Haldane discusses the possibilities of glandular physiology, endocrinology and pharmacology, and sketches a biological utopia in which disease has been abolished, and death is no more to be feared than sleep. His strong message is that biology will bring not one change, but many – I have mentioned only a selection here – and that it is their combination which will be important. The reader can begin to imagine a society in which people are born from ectogenetic vessels, spend their lives under the influence of a range of new drugs, are nourished by an unlimited variety of scientifically produced foods, and pass fearlessly and peacefully towards a painless death. All of this is said to rely on discoveries just over the scientific horizon.

This inspired vision of the biological wonders to come caused a stir. The book passed through five impressions in its first year in print, and sold nearly 15,000 copies in the UK. Across the Atlantic an abridged version appeared in *Century Magazine*.[36] In keeping with the times, one reaction was to accept his scientific but not his social premises, taking issue with Haldane's cheery optimism. Haldane, a lifelong utopian, concluded *Daedalus* with a confident assertion that 'the tendency of applied science is to magnify injustices until they become too intolerable to be borne'.[37] The firmest rebuttal to *Daedalus* on that score was Bertrand Russell's *Icarus, or the Future of Science*, which quickly appeared in the same series.[38]

Russell did not discuss the biological ideas directly, but looked at the existing results of industrialisation. As he said in his essay's opening page, 'I am compelled to fear that science will be used to promote the power of dominant groups rather than to make men happy.' The implication was clear. If the life sciences were potentially so powerful, then their potential for abuse was commensurately great. Russell elaborated this view in a full-length book, *The Scientific Outlook*, published in 1931, and widely taken as a key source for *Brave New World*. In a section of the book on scientific technique in society, he focused on the potential of psychology and biology. In his view, 'men will acquire power to alter themselves, and will inevitably use this power',[39] with the consequence that man 'will tend more and more to view himself as a manufactured product'.

Young Huxley – popular science and admonitory fiction

Russell and Haldane stood on opposite sides of the debate about the impact of science and society, but not simply because they were scientist and philosopher. Julian Huxley, Haldane's lifelong friend, had equally strong credentials as a biologist, but in his work as a populariser ultimately leant toward Russell's position. Huxley, who in the early 1920s was professor of zoology at King's College London, was one of the few scientists who wrote popular articles, as his grandfather had done. His career in print began in 1921 when his work on induced metamorphosis in the axolotl was published.[40] The newspapers reported that 'Young Huxley has discovered the Elixir of Life',[41] and he tried to limit the damage to his reputation by setting the record straight. The essay which resulted began a long career of writing for the lay reader which included the massive *Science of Life* I quoted from earlier, written at H.G. Wells's instigation but mostly from Huxley's pen.

Huxley's writing career suggests that there was an appetite among the educated public for interpretations of biological discoveries in a period when few others were writing authoritatively about the science. Huxley's efforts, always lavishly experimentally detailed, sometimes severely technical, although mitigated by the elegance of his prose, assumed a genuine desire to understand the scientific and technical details of new work. The essays appeared in a range of up-market magazines, which limited their circulation, although they had an enduring life as collections. The first of these, *Essays in Popular Science*, appeared in 1926, went through two printings in a cheaper edition in 1929 and 1933, and was an early Pelican paperback in 1937.

In the essays, a key premise is the belief, shared with Haldane, in the potential significance of biology. But, as I have already suggested, Huxley tended more to Russell's view of the likely uses of biological capabilities than to Haldane's. In his very first essay, 'The control of the life cycle', which originally appeared in the *English Review* in 1921, he concluded that

> The greatest event of the nineteenth century was the revolution caused by man's sudden stride to mastery over inorganic nature. The twentieth will see another such revolution caused by another step forward in mastery, but this time mastery over organic nature. The revolution is even likely to be greater than its predecessor; for the stuff that will be controlled is the basis of our thoughts and emotions and very existence.[42]

Like Haldane a few years later, Huxley thus formulated explicitly what had been implicit in so many fictions of science before his time. This perception, at the root of both Haldane's hearty optimism and Russell's patrician pessimism, this time merely prompts a series of rhetorical questions:

> Are we prepared for such a change? Can we truthfully say that we have used well the power put into our hands by the discoveries of the last century? Or that we are in the least likely to use well any further power such as imagination sees maturing in the womb of discovery today?

Huxley's liberal humanist answer is that the new knowledge will demand a society in which reason prevails, but the answer lacks conviction. Elsewhere, he gives more concrete examples of the kind of powers he has in mind. In genetics for example, he wrote, in a piece first published in *Discovery*,

> A glimpse has been revealed of the kind of machinery that is at work. This, in chemistry and physics, led rapidly to an enormous extension of our

control over lifeless matter . . . It is inevitable that the same increase of control will follow in the realm of living things; only here, because of the greater complexity of the problems, and the greater labour involved in keeping and breeding animals for experiment, it will take longer to achieve an equal result.[43]

Later in the same essay he repeats the point in a discussion of mutations:

At present we do not know how to produce mutations, but *the belief that we shall eventually be able to do so underlies our work,* and once we have discovered the way, our knowledge of the laws of heredity will enable us to build up improved races of animals and plants as easily as the chemist now builds up every sort and kind of substance in his laboratory.[44]

The use of 'races' gave a clear signal to the reader that Huxley the eugenist would welcome the use of this new applied biology for the improvement of human beings, though this time he stopped short of spelling this out.

The theme of biological control echoes time after time in Huxley's essays, usually as a coda to a more sober discussion of contemporary research.[45] In his non-fiction, he left the potential consequences for the reader to imagine. But in his sole piece of fiction, the short story 'The Tissue Culture King', he tried his hand at a satirical treatment of some of those consequences. The story, which appeared in the *Cornhill Magazine* – then still edited by Julian's father – and in the *Yale Review* in 1926, was an effort to put across in a less restrained way, his misgivings about the uses of research. His popularisation, written with enough scientific rigour to satisfy his colleagues, was mainly a matter of describing technical details gracefully. Huxley's admonitions came at the end of quite long and complex narratives, and were rather general. In fiction he was free of some of these restraints, though the story still has a rather didactic air.

The tale opens with a party of white explorers in Africa entering uncharted country, and startled to observe a two-headed toad cross their path. Pressing on, they soon see a variety of other freaks, giants and dwarves. The sightings continue during their journey across country after their capture by a band of giant natives. In their central village, they discover a white man, Hascombe, 'lately research worker at Middlesex Hospital, now religious adviser to His Majesty King Mgobe'.

Hascombe reveals that he was a medical student who progressed to research, and pursued work on parasitic protozoa, tissue culture, cancer and developmental physiology. On an African field trip, studying sleeping sickness, he was captured by the tribe he now lives with, and held prisoner. To save himself – observing the role which blood played in their religion – he began to make medical demonstrations, beguiling them with the histology of blood cells. With the aid of a high-ranking villager, who recognised that knowledge was power, he set about modifying the native religion, having secured a laboratory and the help of slaves and priests. The four elements of their worship, Hascombe finds, are veneration of the king, animal fetishism, sex, and ancestor worship. 'Hascombe reflected on these facts. Tissue culture; experimental embryology; endocrine treatment; artificial parthenogenesis. He laughed and said to himself: "Well, at least I can try, and it ought to be amusing".'[46] Hascombe's laboratory facilities include 'stables with dozens of cattle and sheep, and a sort of experimental ward for human beings'. Huxley thus invokes the spirit of nineteenth-century fictions but this is a parable of modern, applied science, with state support. His reference to twentieth-century research is detailed, if a little heavy-handed. For example, Hascombe refers to his 'Institute of Religious Tissue Culture'. The narrator explains:

> My mind went back to a day in 1918 when I had been taken by a biological friend in New York to see the famous Rockefeller Institute; and at the words tissue culture I saw again before me Dr Alexis Carrel and troops of white-garbed American girls making cultures, sterilizing, microscopising [*sic*], incubating and the rest of it.

Hascombe's greatest success is in making tissue cultures from the king; these come to be used as extensions of the priest-monarch in the ritual life of the tribe. He describes the schemes for selling cultures of the king to the tribespeople, and to encourage the young women to enrol as Sisters of the Sacred Tissue, but his point is a serious one: 'Thus did Bugala [Hascombe's native ally] effect a reformation in the national religion, enthrone himself as the most important personage in the country, and entrench applied science and Hascombe firmly in the organisation of the state.'

Hascombe extends the same treatment to other aspects of the faith: making cultures from ageing relatives before they die; raising a sacred bodyguard of pituitary giants and dwarfs; and instigating a caste of obese virgins. These give rise to further experiments. Hascombe relates:

The obese virgins have set me a problem which I confess I have not yet solved. Like all races who set great store by sexual enjoyment, these people have a correspondingly exaggerated reverence for virginity. It therefore occurred to me that if I could apply Jacques Loeb's great discovery of artificial parthenogenesis to man, or, to be precise, to these young ladies, I should be able to grow a race of vestals, self-reproducing, yet ever virgin.

Huxley thus succeeds in devising applications for both of the discoveries which raised such a furore a generation before. One other authentic line of work mentioned is in embryology, which Hascombe uses to produce animal monstrosities. This was 'of course, done years ago in newts by Spemann and fish by Stockard; and I have merely applied the mass-production methods of Mr Ford to their results'. Hascombe is a complete scientific opportunist, tailoring his work to his captors' beliefs. He no longer wants to escape, the work is so promising. New experiments will lead, he hopes, to mass telepathy. When he and the narrator do break free, using a hypnotic technique of Hascombe's devising on the villagers, the scientist is called back by the same telepathic means, and only the storyteller makes it back to England. Once there, he spells out the moral for the reader:

Dr Hascombe attained to an unsurpassed power in a number of the applications of science – but *to what end did all this power serve?* It is the merest cant and twaddle to go on asserting as most of our press and people continue to do, that increase of scientific knowledge and power must in itself be good . . . I ask [the great public] what they propose to do with the power which is gradually being accumulated for them by the labours of those who labour because they like power, or because they want to find out the truth about how things work.[47]

Huxley's story thus served as a contribution to the developing debate about relations between knowledge and power. And his scientifically informed picture of future biological possibilities and their uses in his microcosmic African society helped prolong the public life of Carrel and Loeb's discoveries, while translating their implications into a new post-war language, in which questioning the equation of scientific advance with social progress was no longer confined to the pages of the *Daily Herald.*

Haldane and Huxley's essays were mainly for the intelligentsia – this was before Haldane's great days as a columnist for the *Daily Worker* – for an audience of a few thousand. But as well as diverting the readers of the *Cornhill Magazine* and the *Yale Review*, 'The Tissue Culture King' also appeared in a magazine which the vast majority of the readers of these journals had probably never heard of, *Amazing Stories*. Out of sight of the leaders of respectable opinion, another area of popular writing which focused on scientific developments was burgeoning. Popular science by elite figures like the two biologists had its audience, but it also furnished ideas for stories read by many more, stories catering to the rapidly expanding market for science fiction.

Pulp fiction – Faust reborn

Amazing Stories, founded by Hugo Gernsback in 1926, was one of the first fiction magazines entirely devoted to science fiction. Printed on pulp paper, it sold cheaply and well, and was soon established as a flourishing offshoot of the much larger pulp magazine market. Look at the August 1927 issue of *Amazing Stories*, in which Huxley's story appeared, and it is not at all out of place, though less flamboyant than the rest. Like the typical Gernsback story, it turned on a number of ideas plausibly derived from contemporary scientific knowledge. Gernsback, who began his career as editor of popular magazines about radio, electrical engineering and mechanics, insisted on 'hard science' when he turned to fiction. His magazine thus offers another source for following responses to contemporary scientific developments. Fiction in *Amazing Stories* and similar magazines frequently contained large chunks of scientific exposition, woven more or less skilfully into the narrative. This made the links between science and imaginative responses to it unusually explicit.[48]

Those responses were generally positive. Gernsback was a famous enthusiast for all kinds of technical innovation, and the overall tone of the stories he published tended toward nineteenth-century optimism. As Brian Stableford suggests, 'Gernsback originally considered scientifiction (as he called it at first) to be an implicitly utopian species of literature, one of whose main functions was to herald a new technological Golden Age', and most of the stories bore this out.[49] Nevertheless, here and there in some of the early Gernsback stories are to be found echoes of the Faustian tradition, strains of antimechanism and a perpetuation of negative stereotypes of the scientist. Most of the stories which take this turn are about biology.

As a review of *Amazing Stories* from 1926 to 1929 – when Gernsback lost control of the title – shows, deviation from the party line was allowed if the story was good enough, and this kind of inspiration most often stemmed from the life sciences.[50] Huxley is a case in point, but others contented themselves with updating the older images of biological experiment, using familiar stereotypes but taking account of more recent scientific developments.[51]

For example, the Faustian theme is the basis of the story 'The Malignant Entity', from June 1926, in which a Professor Townsend is striving to create a living organism from inert matter.[52] When he succeeds, he is faced with a soulless something, and his partly digested skeleton is found in the laboratory, consumed by his creation. There is, of course, nothing at all new in such a story. Even when symbolically more sophisticated tales such as *R.U.R.* had been widely circulated, the old images continued to give life to newly written stories. The new elements are assimilated and blended with the old in popular literature, whose authors continually rework and update the established plotlines and characters, confident of their enduring appeal.

Thus the solitary scientist meeting his individual fate reappears again and again, but rarely with any effect on the wider society. A similar story, with a yet more explicitly Faustian moral, is 'The Evolutionary Monstrosity', which appeared in *Amazing Stories Quarterly* in 1929.[53] This time, an attempt to speed up evolution gives rise to the monstrosity of the title. It becomes apparent that the monstrosity created is in fact a transformation of one of the experimenters, who thus learns that 'Man dare not tamper with God's plan of a slow uplift for all humanity'. Another story with a distinctly Victorian flavour is 'The Head', from the previous year, in which the inhuman Dr Leeson performs diabolical experiments with a severed head, kept alive with a mechanical heart and artificial blood. The head is conscious and suffering, but he refuses to put it out of its misery. Like Moreau, he regards a human being 'only as an object for his probing knife'.[54]

Sometimes the unfeeling scientist would appear in an explicitly antimechanist narrative. Čapek's nineteenth-century materialist attempting to create life in order to prove God unnecessary reappears in 'Blasphemer's Plateau' from 1926, surrounded by much up-to-date scientific apparatus. The narrator goes to visit an old friend, who lives in an isolated house on the outskirts of a gossip-ridden village. His friend, a biologist, has turned in later life to trying to prove 'that spiritual immortality does not exist . . . and that goes for Resurrection too' by creating living things from inorganic matter. He has created cells by irradiating chemicals, and has grown fertilised ova and, as he tells it, we have gone beyond even this. We know the composition of the eggs and ova of more than fifty varieties of organism. We have duplicated them successfully and activated them with vibratory impulses equivalent to the fertilization and germination processes of 'Nature'. We control the sex at will by limitation of the chromosomes of the primary cell.[55]

As soon as they have mastered the construction of tissues, the biologist and his assistant plan to 'use the human ovum now in Johnson's incubator, *and artificially create a human being!*' (original emphasis). Their visitor grows more and more uneasy: 'never had he sensed so sinister an aura as that which surrounded the quiet-voiced, mild-mannered scientist who droned of his hopes and accomplishments'. His growing repugnance leads him to try and stop the work. He fails, and is turned into an idiot by the biologist, but retains his sense just long enough to see the scientist and his aides succumb to a prolific cancer, induced by all that radiation. They are destroyed just as they are about to confer intelligence on the human baby in the incubator.

Some stories played around with the image of ectogenesis as elaborated by Haldane. In 'The Machine Man of Ardathia', from 1927, the narrator is visited by a highly evolved man from the future, who lives inside a glass cylinder – a giant test-tube. The creature explains how his race developed.

The cell from which we are to develop is created synthetically. It is fertilized by means of a ray and then put into a cylinder. As the embryo develops the various tubes and mechanical devices are introduced into the body by our mechanics and become an integral part of it. When the young Ardathian is born, he does not leave the case in which he has developed.[56]

The Ardathians live a life of unemotional contemplation, which lasts 1,500 years. The narrator can see how it all began, recalling a Russian scientist reported in the press as keeping a dog alive with an artificial heart, and that 'I read an article in one of our current magazines telling how a Vienna surgeon was hatching out rabbits and pigs in ecto-genetic incubators'. He also remembers that he had 'heard tell of chicken hearts being kept alive in special containers which protect them from their normal environment'. The Ardathian compliments him on the advances already made along the path of future evolution. But as the details of Ardathian life are spelled out, the narrator is horror-struck:

I began to get an inkling of what the Ardathian meant when it alluded to itself as a Machine Man. The appalling story of man's final evolution into a controlling center that directed a mechanical body awoke something akin to fear in my heart. If it were true, what of the soul, spirit, God . . .[57]

These stories are often crude, of course, and are of minimal literary interest. They most often follow the mad scientist variation of the knowledge narrative summarised by Andrew Tudor in his study of film. All their ideas are drawn from others – Shelley or Wells, Čapek or Haldane. Yet they remind us once again that the major motifs of *Faust, Frankenstein, Moreau* or *R.U.R.* have a life in popular print culture beyond reproduction of the original tales. They are constantly appropriated in new stories, retold in modern dress.[58] Nor was this collective retelling confined to *Amazing Stories*. The prolific and popular Edgar Rice Burroughs, for example, published a number of stories in this vein, including *The Monster Men*, a sub-Wellsian novel first published in 1929.[59]

In some cases, the new science fiction stories were simultaneously the vehicle for popularising a new scientific discovery and for linking it directly to the negative images originally attached to older findings. A case in point is H.J. Müller's successful production of artificial mutations in fruit flies by exposing them to X-rays, first reported in 1927. This was one piece of biology which did make the newspapers,[60] and was quickly taken up in fiction. In 1934, for instance, a novel appeared based on *Moreau* but using genuine mutation techniques in place of vivisection.[61] Müller's work also furnished the premise for a striking story which appeared in the pulp magazine *Wonder Stories* in 1936. 'The Master of the Genes', by a regular author for the pulps, Edmund Hamilton, opens with a geneticist recruiting two Americans for protection from the inhabitants of the Amazonian jungle village where he is working. The scientist, Dr Alascia, is studying a rash of abnormal births among the Indian villagers, and as the number of monstrosities grows the Indians are becoming restive. His would-be protectors soon discover that he is causing the mutations himself by bombarding the villagers with X-rays. The source of this gruesome inspiration is acknowledged in confrontation with one of the Americans, who tells Alascia:

In biological books in your library . . . I read how Müller of the University of Texas discovered that X-rays would disturb and damage the genes, and how he had produced monsters in that way among fruit flies. I remembered that huge X-ray machine in your laboratory and saw that you'd been repeating Müller's experiments with humans![62]

Dr Alascia dismisses their anger as 'the usual foggy humanitarian sentimentalities', but soon learns that his own daughter has become pregnant while in the village. She discovers her child-to-be will also be deformed, and commits suicide, following which the doctor too despairs, and is killed by the Indians.[63] But his earlier words pose a question for the future of his science:

> Why should science know so much of the gene system of fruit flies and so little of the human gene system? Simply because sentiment forbids experimenting with humans, I determined to disregard that sentiment to make this place a laboratory for gene experiments on humans . . . Some day this data I have amassed . . . will help geneticists of the future build a better human race.

With hindsight, Hamilton's story appears doubly prescient, about eugenic fears attaching to genetics and about post-Nazi concerns over human experimentation. At the time, it was just one more tale in the pulps.[64] In these stories it is the collective message that matters, not the detail of any individual text, however striking. But there is one final text to discuss here which, like *Frankenstein*, seems to sum up the discussion of a whole milieu in a work of fiction which, well crafted and timely, attains near mythic status. That text, of course, is Aldous Huxley's *Brave New World*.

Babies in bottles – biology and dystopia

The elements of Huxley's novel were not new. The book wove together the strands of future-oriented discussion which had appeared through the 1920s. We can get an idea of the features of *Brave New World* which gave the images of that book their peculiar resonance by looking briefly at an opposing synthesis, published just two years before. This was the Earl of Birkenhead's *Life in the Year 2030*, a projection of the world a hundred years in the future.

Birkenhead's book was one of a number of similar efforts, most now obscure. A former Tory Lord Chancellor and Secretary of State for India, Birkenhead wrote many popular books as well as more scholarly works on law and politics. Clarke suggests that the fact that he wrote a futurological book was 'an important sign that . . . the future had become a matter of popular interest'.[65] Many of the biological ideas he took more or less unchanged from Haldane, although the general proposition that biology had special significance was becoming a commonplace.[66] Birkenhead's own formulation was that 'Any developments in physics and chemistry which can be reasonably predicted to occur before 2030, can do no more than alter the accidentals of human existence. In biology, however, developments may be predicted which will change the whole nature of life as we experience it today.'[67]

Julian Huxley, after making the same point in his essay on 'The control of the life cycle', had gone on to ask whether people were ready for such a change, whether the power to alter the conditions of biological existence could be used wisely and well. Birkenhead's answer put him among the optimists. His was a book written by one whom the world had treated well.

He foresaw prodigious advances in medicine and surgery, the abolition of epidemic disease, painless childbirth, life prolonged through rejuvenation, and synthetic food that would make agricultural labour obsolete. In all his catalogue of innovations there is only one about which he seems in the least ambivalent, only one point where optimism falters for a moment. This is in the discussion of ectogenesis. To be sure, he sees it as inevitable, and necessary for the state to ensure adequate supplies of good-quality citizens. Birkenhead feared the eugenic impasse, believing that 'the indiscriminate increase of the most useless type of citizen, accompanied by the voluntary sterilization of the best type, is the greatest menace which threatens our civilization'.[68] Ectogenesis offers a way out, an acceptable way of regulating births which obviates the 'human stud farm' methods of the 1920s eugenicists. So although he predicts much religious opposition, Birkenhead regards the adoption of ectogenesis as a foregone conclusion. It will have the added appeal of the liberation of women: 'One day [women] will accept ectogenesis as the price for a freedom they have never yet achieved in history.'

Yet in spite of this confidence, when ectogenesis is first introduced a note of uneasiness creeps in. He apologises for burdening the reader with the topic:

> The possibility of ectogenetic children will naturally arouse the fiercest antagonism. Religious bodies of many different creeds will rally their adherents to fight such a fundamental biological intervention. In fact the mere mention of its possibility here may strike the reader as gratuitously affronting.

However, the implications must be considered because 'the thing is possible; and since it is possible, it is certain that scientists will be deterred by no persecution from straining after attainment'.[69] His uneasiness enters because he realises that, as well as the 'shattering' social effects of a separation of reproduction from marriage, ectogenesis might enable the 'character of the future inhabitants of any state [to] be determined by the government which happened temporarily to enjoy power'. He expands on what this might mean:

> By regulating the choice of ectogenetic parent of the next generation, the Cabinet of the future could breed a nation of industrious dullards, or leaven the population with fifty thousand irresponsible, if gifted mural painters . . . If it were possible to breed a race of strong healthy creatures, swift and ductile in infinite drudgery, yet lacking in ambition, what ruling class could resist the temptation?

This particular member of the ruling class certainly seems as unhappy about the consequence of his eugenic and determinist beliefs as the leader writer of the *Daily Herald* in 1912, whose words he echoes, was about the prospects for creating life:

> The ectogenetic slave of the future would not feel his bonds. Every impulse which makes slavery degrading and irksome to ordinary humanity would be removed from his mental equipment. His only happiness would be in his task; he would be the exact counterpart of the worker bee.[70]

However, this glimpse of a less desirable future, so out of keeping with the rest of the book, is soon glossed over. Birkenhead's future society will be so ordered that such things never happen. When limitless atomic energy and synthetic food furnish a life of ease, 'production will become so cheap and . . . wealth will accumulate to such an extent that the ectogenetic robot will never be needed'. It was Aldous Huxley's task to explore this glimpse of a biologically stratified society in more detail.

In *Brave New World*, published in 1932, Huxley expanded Birkenhead's fleeting vision into a full-length novel. He based his society on ectogenesis and conditioning, although also incorporating many other possible innovations discussed in the 1920s. But the core of his vision, and the main vehicle for the development of his own views on the prospects for scientific and social progress, was biological speculation. In Huxley's book, the biological bases of human social life are transformed by the whole range of techniques then thought to be on offer: hereditary selection, chemical and psychological conditioning, depressive and euphoric drugs, enforced contraception, prolongation of youth, and euthanasia. The emphasis on biology was deliberate, born of Huxley's wish to portray 'the advancement of science as it affects human individuals'. In the foreword to the 1950 edition of *Brave New World*, he explained how, in the book, 'the triumphs of physics, chemistry and engineering are tacitly taken for granted. The only scientific advances ... specifically described are those involving the application to human beings of the results of future research in biology, physiology, and psychology.'[71]

He shared the view of his brother, and of Haldane and Birkenhead, that 'It is only by means of the sciences of life that the quality of life can be radically changed.' This was the perception which informed the whole book. Huxley combined ectogenetic birth *à la* Haldane with mass production and made them the basis of a completely imagined dystopian society.[72] The link between ectogenesis and the production line is 'Bokanovsky's process', by which a large number of identical individuals can be produced from one fertilised egg. This underlines the suppression of individuality in the interests of social stability (significantly, the process is only applied to the lowest types – Gammas, Deltas and Epsilons), and Huxley repeatedly emphasises its importance. 'I see you don't like our Bokanovsky Groups,' Mustapha Mond tells the Savage, 'but, I assure you, they're the foundation on which everything else is built.'[73]

This reproduction of identical humans with capabilities restricted to menial tasks recalls Čapek. Bokanovsky's process makes explicit a possibility that Birkenhead also foresaw – the use of ectogenesis to make robots; 'the principle of mass production at last applied to biology,' the Director of Hatcheries and Conditioning tells his students, echoing *R.U.R.* In *Brave New World*, however, the production line is taken one step further. Instead of the robots gradually taking over from humans, the humans themselves *are* robots, and have no prospect of transcending their inbuilt limitations. The final irony of Huxley's scenario is that the labour of the lower classes is technically unnecessary but must be perpetuated in the interests of stability. Further technical innovation, Birkenhead's saving grace, is suppressed in Huxley's world. Mustapha Mond explains:

We could synthesize every morsel of food if we wanted to. But we don't. We prefer to keep a third of our population on the land. For their own sakes – because it takes *longer* to get food out of the land than out of a factory.[74]

This is ultimately in the labourers' best interests; it would be 'sheer cruelty to afflict them with excessive leisure'. Besides:

We have our stability to think of [Mond continues]. We don't want change. Every change is a menace to stability. That's another reason why we're so chary of applying new inventions. Every new discovery in pure science is potentially subversive; even science must sometimes be treated as a possible enemy. Yes, even science.

The book thus offers a society in which the ultimate result of applied science is to bring development to a halt – a direct contradiction of the actual experience of modernity. But Huxley's near-echo of Haldane is intended for his contemporary readers, and they find this renewed suggestion of science's subversive potential in what is at heart a profoundly pessimistic book.[75] Huxley's future society is deeply unattractive, but is described in unprecedentedly realistic detail. The famous opening chapters describing in cinematic mode a conducted tour of the Hatcheries – the ectogenetic production line – are still an arresting succession of images.[76] But it is not only the quality of the writing which carries conviction, it is also the combination of ideas never before presented in this form.

The result was a book which had 'a devastating effect on the intelligentsia of the Western Hemisphere' in their thinking about the future.[77] And not only on the intelligentsia; from its first publication, the image of this particular Brave New World became one of the dominant motifs in all subsequent discussion of biological discovery – whether scholarly or popular. The Shakespearean title and its associations have entered the language in much the same way as 'Frankenstein', even without benefit of translation into a form as compelling as the *Frankenstein* films. The number of citations of or allusions to *Brave New World* in later texts is enormous. But it is still hard to do justice to the cumulative impact of the book. It looms large over a whole area of public debate, shaping reactions to and evaluations of a wide range of biomedical research.

The book was widely discussed when it was first published. Huxley was already an established novelist, with works like *Point Counter Point*, and *Brave New World* was reviewed in the daily press and the major literary journals.

Many reviewers found *Brave New World* lacking as a novel, or were opposed to its view of progress, and a variety of critiques appeared. These ranged from accusations of poor writing to denunciations of the book as a naive piece of propaganda.[78] But there were also favourable responses. Rebecca West wrote in the *Daily Telegraph* that it was 'almost certainly one of the half-dozen most important books that have been published since the war'. West also inaugurated discussion of a point which preoccupied many later writers: how close are the techniques Huxley described? Many seemed to assume that the biology and psychology of *Brave New World* were far in the future; West was curious to know if this was true:

> If one has a complaint to make against him [Huxley] it is that he does not explain . . . how much solid justification he has for his horrid visions. It would add to the reader's interest if he knew that when Mr. Huxley depicts the human race as abandoning its viviparous habits and propagating by means of germ cells surgically removed from the body and fertilized in laboratories . . . he is writing of a possibility that biologists see not more remotely than, let us say, Leonardo da Vinci saw the aeroplane.[79]

An answer came quickly from Joseph Needham, in May 1932, and it remains one of the most confident and the most startling. Needham, then a prominent young embryologist and a former colleague of Haldane and Gowland Hopkins, saw the book as about the powers of an autocratic dictatorship 'given the resources of a really advanced biological engineering'. He suggested that those who found Huxley's vision distasteful would say the biology must be wrong. Not so, said Needham:

> What gives the biologist a sardonic smile as he reads it is the fact that *the biology is perfectly right*, and Mr. Huxley has included nothing in his book but what might be regarded as legitimate extrapolation from the knowledge and power we already have. Successful experiments are even now being made in the cultivation of embryos of small mammals *in vitro*, and one of the most horrible of Mr Huxley's predictions, the production of numerous low-grade workers of precisely identical genetic constitution from one egg, is perfectly possible.[80]

This kind of endorsement has repeatedly helped to reinforce the influence of *Brave New World*. Genuine ectogenesis is of course still far from being achieved today, but as with the 'creation of life' it became established in the popular commentary as an implicit goal of a particular line of research.

This process in fact began very early in the history of *Brave New World*. When the book was published, Gregory Pincus was establishing himself at Harvard as an authority on sexual physiology, and he had begun experimenting on rabbit eggs. In 1934, he announced that he had achieved *in vitro* fertilisation in rabbits, and was depicted in the *New York Times* as a real-life Bokanovsky. In reality, he was stimulating the eggs to develop without normal fertilisation, just as Loeb had done with sea urchins, and he achieved similar notoriety. Later papers from Pincus led to further reports in the *New York Times*, *Time* and *Newsweek*, and a swingeing attack on his work in *Collier's* magazine, which saw it as an assault on the American male.[81] Pincus was subsequently denied tenure at Harvard, and had to pursue his research outside the universities.

He was just one of what became a ready supply of scientists prepared to speak to the goal of *in vitro* fertilisation. And whenever they did, old associations clung to their predictions about new science. Just as the roots of Huxley's vision can be seen in the *Daily Herald*'s sketch of the consequences of 'creating life' in 1912, so today's practice of *in vitro* fertilisation is represented as the creation of life in the test-tube.

Despite Needham's comments, and Pincus's research, *Brave New World* and other books of this period about biological futures were still more often taken as distant prophecies than as suggestions of immediate prospects. Pincus was the main exception to the generalisation that the work actually being done in laboratories at this time rarely made news. The speculations of Haldane or the Huxleys, while they were related to general debates about science in which they were then involved, were more important in setting the scene for later discussions of biology than for expressing perceptions of biology as it then stood. Popular images of biomedical science were still strongly influenced by the past successes of experimental investigation: in plant and animal breeding, in control of agricultural pests and, above all, in medicine. Bestselling books like Paul de Kruif's *Microbe Hunters* – an exuberant piece of hagiography – helped reinforce scientific medicine's image as a modern cure-all, and the continuing discoveries of new drugs and vitamins seemed to fulfil this promise.[82] Books like this form an important genre of popular writing in which, as the cover of a wartime paperback edition put it, 'heroes of science battle disease'. Similar works continued to appear into the early 1960s, exemplified by Ritchie Calder's *The Life Savers* of 1961, which relates victories in a war against the 'death-dealing enemies of mankind'.[83]

These books show the persistence of the contradiction in outlook which was evident in responses to vivisection, or to Loeb and Carrel's work – enthusiastic approval of the medical benefits of biological research alongside disquiet about its other possible applications. Since the time of *Brave New World*, this contradiction has appeared more and more often in different attitudes to research already done and work yet to be accomplished. Reviews of past achievements emphasise the benefits they have brought, while visions of the future are much more likely to be fearful. There are exceptions of course – Haldane for one – as well as critical evaluations of successes like the emphasis on the role of improved medical care in population increase. Perhaps it is more accurate to say that the full impact of pessimistic views about the implications of biological advances was first felt in discussions of the future, in the 1920s and 1930s, but since then it has affected discussion of contemporary developments more strongly.

One reason is that laboratory research in biology now appears to have moved much nearer to the transforming capabilities envisaged by Huxley. In the popular view, the science is ever catching up with the fiction. The goal the writer outlines with a single imaginative leap is taken to be the ultimate destination for many small steps taken by the scientist. Much of the later discussion of new biological discoveries has turned on interpretations of the relation of the work involved to these kinds of goals. More recently, the tendency to conflate scientific and fictional agendas has been more actively resisted by researchers. But there have usually been scientists of repute who have endorsed it, as well, for their own personal or institutional reasons.

More significantly, perhaps, some researchers and orchestrators of research have been directly inspired by the vision of biological control. Thus, in the same year as *Brave New World* appeared, we find Sir Walter Morley Fletcher in an address to a conference on biology and education suggesting that 'physical science had greatly changed the world, but biology would change it far more, for it would change man himself, not merely the environment'.[84] And as secretary of the Medical Research Council, who better to ensure that the possibilities of experimental biology would be realised?

Generally, though, there were few indications before World War II that the biological prophecies the public was hearing were likely to come true in the near future. It was characteristic of the period that the work most widely hailed by the public was that of Alexis Carrel, whose most important achievements dated from before World War I. In 1935, he achieved renewed fame as the builder of a 'glass heart' for maintenance of isolated organs outside the body. Much of the interest arose because Carrel, the Nobel Laureate, collaborated with Charles Lindbergh, a huge hero in the US for his flying exploits and also a talented mechanical engineer. Newspaper reports played up Lindbergh's part in the device.[85] Nevertheless, there was also some emphasis on the potential of the 'artificial heart' for the study of disease, and *Time* speculated about the prospects opened up by cultivation of cat and chicken ovaries in the machine. It was suggested that 'laboratory technicians conceivably might someday fertilize and incubate such motherless eggs to produce chicks or kittens'.[86] Carrel spoke of his hopes of storing organs for transplantation, or of placing damaged organs in a culture chamber, which led to his successes being 'recklessly exaggerated by the sensational papers, some of which went so far as to suggest that he planned to propagate human babies in vitro, or to keep an isolated human brain alive and thinking'.[87]

But Carrel's work of the 1930s was exceptional, both in its visibility and because it was out of the main line of development of biology. For in the 1930s, largely unheeded by press and public, the conditions for the rise of molecular biology after World War II were being established. The crucial techniques – crystallography, chromatography, radioactive tracers, the ultracentrifuge – were all being developed. Patterns of research funding were changing. During the 1930s, the Rockefeller Foundation, Carrel's sponsor, pledged a large sum to an expanded programme of fundamental biology. This programme, at the instigation of Warren Weaver, was to be concerned with rigorous, physico-chemical, quantitative basic research, related only obliquely to medicine and harking back strongly to the spirit of Jacques Loeb's programme for biology. It was summed up in Weaver's new term – molecular biology.[88]

In the early years, this type of research did not translate readily into the terms of popular discussion. The work was abstruse, complex and far removed from everyday experience. Consider, for example, perhaps the most striking pre-war achievement of the Rockefeller programme, the crystallisation of tobacco mosaic virus by Stanley. This happened in 1935, just when Carrel's glass heart caught the journalists' eye, but whereas Carrel's work was the fruit of a research programme going back thirty years or more, Stanley's was an indication of the direction the new biology would take after the war. It confirmed the implications of Sumner's earlier crystallisation of the enzyme urease in 1926: that biologically active substances had specific, regular structures. Perhaps, then, they were not so complex after all. Here began 'the replacement of the colloidal conception of life by the structural conception'.[89] Here, too, began the path to an ability to manipulate life far beyond anything previously foreseen.

By and large, this implication was lost on the wider world. Press reports prompted by Stanley's announcement interpreted his results as narrowing the gap, or blurring the distinction between life and non-life, or casting light on the origin of life. But investigation of viruses, forms of life so lowly they were hardly organisms, was not seen as relevant to the understanding of living creatures of a higher order.[90] So while the ideas and images which would feed into post-war debates about biological control can easily be seen maturing in the speculative essays and popular literature of the inter-war period, the changes in biology that would soon seem to promise the powers the imaginative writers foresaw were beginning to take hold, but were as yet unnoticed. After the war the two began to come together in new ways, as we shall see in the next chapter.

Chapter Six

Time-bombing Our Descendants

Must we geneticists become bacteriologists, physiological chemists
and physicists, simultaneously with being zoologists and botanists?
Let us hope so.

H.J. Müller, 1921

Nuclear hopes and fears

In the aftermath of World War II science took on a heightened significance,
both for the public and for the state. The war, and especially the Manhattan
Project to build the A-bomb, crystallised pre-war trends in scientific organisation. And the destruction of many European centres of research combined
with a newly strengthened relationship between scientists and government in
the United States to complete a shift in the scientific centre of gravity toward
North America. In the US, the nuclear age was to be the time of 'big science';
of centralised research organisation, concentration of work in major laboratories and a steadily expanding science budget. Other industrialised countries
followed suit. Across the Western world, state support for science was available on a hitherto undreamt of scale.[1]

To begin with, the political advantages from the 'physicists' war' were most
obvious in the big sciences of high energy physics and radio astronomy.
Biology's financial demands were small by comparison, and a general increase
in funds quickly embraced research in the life sciences as well. National
funding agencies like the newly chartered National Science Foundation in the
US began to take up some of the burden of support for projects previously
paid for by private promoters of biology like the Rockefeller Foundation. In

Britain, for example, the Medical Research Council took over the Rockefeller-backed research of Max Perutz and John Kendrew on protein structures at the Cavendish Laboratory in Cambridge. Later, as the financial appetite of biologists grew, the general increase in research budgets more than satisfied them, at least until the mid-1960s. In the US, for example, spending by the National Institutes of Health rose from $81 million to over $300 million between 1955 and 1959.[2]

Immediately after the war, the Atomic Energy Commission's programme on radiation genetics paid for much American biological research. And here, too, was one of the central areas of public interest in science. For the public, as for politicians, intellectuals and many scientists, the most important scientific product of the war by far was the Bomb.[3]

To understand the impact of atomic weapons on public attitudes to science, a brief historical digression is needed. So far, I have focused almost exclusively on biological science, and reactions to biological science. But the most awesome products of physical science must have a place in any consideration of ideas about science and scientists in the twentieth century. The fullest account is that of the historian of physics Spencer Weart, whose study *Nuclear Fear* is in some ways an inspiration for this one. For he wrote, as he says, not a history of the science, but a history of images, trying to get at the roots of our attitudes to and feelings about nuclear physics. As I am approaching biology, so Weart approaches physics, asking what is new and what is old in the images attached to what appears a quintessentially modern science. His original observation was about the times in the 1960s and 1970s when the debates over nuclear energy were, as he saw it, emotional, and the emotions in play were anger and fear. As he wrote in an earlier paper: 'What if there are feelings about nuclear energy that motivate the debate itself, so that the arguments over politics, values, and scientific claims are the side effects, the symptoms of a deeper conflict?'[4] If so, he suggests, then we need to understand the origin of these feelings. I will summarise his interpretation of the history of nuclear fears, before (in part) disagreeing with the later portions of his story.

Weart begins, not with Hiroshima and Nagasaki, but at the turn of the century. The discovery of X-rays in 1895 very quickly caught the public fancy. When radioactivity was discovered, it was not at first recognised as separate from X-rays and radio waves. But after 1900 radium, with its very intense radioactivity, began to arouse wider interest. In 1901, Ernest Rutherford and Frederick Soddy discovered what was actually happening during radioactive decay. One atom was changing into another. Weart quotes one of the discoverers:

'I was overwhelmed,' Soddy recalled, 'with something greater than joy – I cannot very well express it – a kind of exultation.' He blurted out, 'Rutherford, this is transmutation!'

'For Mike's sake Soddy,' his colleague replied, 'don't call it *transmutation*. They'll have our heads off as alchemists.'[5]

This idea of a new alchemy was widely taken up, by scientists as well as in the press. So the nuclear story began by harking back to one of the fundamental sources of imagery framing our view of the getting of knowledge, by alchemists, then by scientists. And transmutation, as Weart emphasises, was already a highly charged symbol within the context of this imagery:

The extent of the danger is particularly shown in the symbol of transmutation. In the crucible of the alchemist, substances were said to die and be reborn, undergoing a descent into corruption and putrefaction before they could be transmuted; this is a symbol of the agonising descent into darkness that is necessary for psychological or spiritual transformations. In short, there is abundant evidence that the symbol of transmutation calls to mind the great theme of death and resurrection.[6]

Looking more widely at myths of cosmic and social decay, of Armageddon and renewal, he concludes that 'there is reason to believe that the moment one brings up the idea of transmutation one must deal with ancient associations involving secret knowledge, personal and social transformation, and shattering dangers ranging right up to the end of the world'.

The notion that fearful consequences might follow the uncovering of radioactivity appeared very quickly, partly encouraged by the physicists. Soddy said that there was so much energy locked up in the atom that the earth must be seen as a munitions pile. The man who could control this energy 'could destroy the earth if he chose'.

As early as 1904, a writer in the (English) *Quarterly Review* related that

It is conceivable that some means may one day be found for inducing radioactive change in elements which are not normally subject to it. Professor Rutherford has playfully suggested to the writer the disquieting idea that, could a proper detonator be discovered, an explosive wave of atomic disintegration might be started through all matter which would transmute the whole mass of the globe into helium or similar gases.[7]

Speculations of this sort – whether 'playful' or serious – quickly became familiar to most literate people, and were often repeated in fiction. The idea was never especially plausible, but neither was it ruled out by what was known of atomic physics.

Aside from the fear of some sort of planetary accident, fears soon arose about the misuse of radioactivity, for military or criminal purposes. H.G. Wells coined the phrase 'atomic bomb' in *The World Set Free*, first published in serial form in 1913–14. He described a nuclear world war of the 1950s in which cities were wiped out, and left contaminated, by induced radioactivity bombs the size of a handbag. It was after Leo Szilard read this novel in 1932 that he became the first scientist to make a serious effort to liberate nuclear energy.

We can see how these fears were borne out, but it is striking to see how early in the century they were first articulated. As I showed in Chapter 4, this was a time of great optimism about science and technology, but when it came to unlocking the energies within the atom there was ample scope for ambivalence. The up-side of the image was a dream of unlimited control over matter, and of our salvation in the face of the already foreseen exhaustion of fossil fuels. Weart quotes Soddy writing in *Harper's Monthly* in 1909 that 'A race which could transmute matter would have little need to earn its bread by the sweat of its brow . . . such a race could transform a desert continent, thaw the frozen poles, and make the whole world one smiling Garden of Eden.' He was also widely quoted as saying that a pint flask of uranium could drive an ocean liner round the world – a notion which was endlessly repeated in the popular press in the following decades.

Even atomic weapons might have their good side. *The World Set Free* offers a vision of a world government, which rises from the ruins of the war and leads to a global regime of social justice and universal prosperity, fuelled by nuclear energy – again a scenario that is now very familiar. The hopes and the fears evoked by nuclear energy were equally huge.

At first, optimism dominated, in keeping with the times. But by the 1930s the physicists were downplaying expectations, on both counts. Leading physicists like Millikan and Rutherford were saying that there was virtually no possibility of extracting the energy bound in the atom. Weart sees this as the first stage in a divergence of attitudes: the coolness of the physicists and a more 'emotional' response in the press. Meanwhile, as with X-rays, medical benefits of radioactivity were widely reported. Radioactive rays were seen as healthful. 'The press reported soberly that radium might well vanquish that paradigm of a dread disease, cancer, and there were more speculative suggestions that radium might halt tuberculosis, make the blind see, and perhaps someday raise the dead.'

On the other hand, the real ill-effects were apparent relatively quickly. Cases of X-rays, and later radioactivity, causing burns, sterility and cancer were widely noted. By the 1920s, their use was regulated in the developed countries, but the associations were still troubling. As we have seen, Müller's discovery of 'transmutation' of genes by X-rays was widely publicised in the latter half of the decade.

Radium was at first seen as possibly generally beneficial − a tonic − and used in patent medicines, but reports of deaths from over-enthusiastic imbibing of such tonics eventually clouded the picture. And radioactive fictions acquired a darker hue. In 1936, Boris Karloff starred in *The Invisible Ray* from Universal. His character, Dr Rukh, builds a radium ray projector that can both destroy cities and save lives. But Rukh becomes contaminated with 'contagious radioactivity', and sets out to avenge himself on the world after his neglected wife leaves him.[8] As with Karloff's screen portrayal of Frankenstein, this was the first of many such characters. However, Weart's judgement is that, on balance, the image of the *medical* uses of radium remained positive in the United States until the late 1930s. He reports a content analysis, based on just the titles of popular journal articles, and finds that even when reservations were at their pre-war peak, positive articles still outnumbered negative ones by two to one. Overall, then, he argues that

> the public found nothing unusually troubling in radioactivity itself. We may hear claims that radioactivity, because it is invisible, malignant, and so on, inevitably arouses irrational fears. No doubt there is some truth in this, but the historical evidence suggests that radioactivity is inherently no more fear-provoking than numerous other invisible, malignant substances connected with industry and medicine (germs and certain chemicals, for example).

By the outbreak of the war, people knew that radiation *could* cause debility, cancer, genetic damage and sterility. But these fears seemed to play little part in reactions to the discovery of nuclear fission, in 1939. That discovery launched another round of speculation about the wonders of a coming nuclear age. What was seen then was essentially a recapitulation of the pattern of the early 1900s, a balance of visions of material prosperity based on cheap energy and forecasts of doom through accident or war.

Then came a real war, the Manhattan Project, and the reality of nuclear weapons. In the aftermath of the first two bombs, there was a vast outpouring of comment. There was a mix of optimism and anxiety, with anxiety now more prominent. There was also, as the cultural historian Paul Boyer stresses, a strong recurrent theme that the world had entered a new age. Undoubtedly there were strong continuities. Weart, again using content analysis, suggests that the American public's views about nuclear energy soon after Hiroshima were pretty much the same as they had been before Hiroshima. However, he argues that the appearance of continuity disguises something more subtle. He suggests that the two bombs had a profound effect on the public mind, causing deep anxiety that tended to be unexpressed. Or, as the conventional wisdom of cultural history would have it, perhaps it was expressed in other ways: in an enhanced fear of radiation, for example, or in the horror films of the 1950s, in which creatures were created or aroused by nuclear weapons.

In time, too, the bombs grew more powerful. The prospect of a political culture in the grip of the Cold War and the advent of strategies for fighting war using hydrogen bombs was a warrant for the most apocalyptic fears, again reflected in novels and the cinema. There were of course many fictions about nuclear war, both in mainstream literature and in the science fiction magazines. There is evidence that those published in the magazines were only a fraction of the total number written. In 1948 John W. Campbell told his readers in *Astounding* that no more 'atom doom' stories would be run, because readers had got the message, and were tired of hearing it. In 1952, the editor of a competing magazine, *Galaxy*, was complaining that 'over 90 per cent of stories submitted still nag away at atomic, hydrogen and bacteriological war, the post-atomic world, reversion to barbarism, mutant children killed because they have only ten toes and fingers instead of twelve . . . Look, fellers, the end isn't here yet.'[9]

Essentially, the rest of Weart's argument is that the various protests which marked nuclear history from the late 1950s on – protests against nuclear fallout from weapons testing and against civil nuclear power, both focusing on radiation dangers – were fuelled by displaced anxieties about the dangers of actual use of nuclear weapons in a final world war.

There is a different reading of this history, though, which builds on the notion that biological effects matter more than physical ones. Apocalypse, mere physical destruction, is rather dull, and dulls the mind in prospect. A flaming finis to civilisation is a possibility to ponder, but can only have a limited grip on the imagination. What grips is a change in the conditions of life, and what would change the conditions of life more than a change in the forms of life itself? This suggests a different explanation for the preoccupation with the effects of radiation (real or imagined) through mutation in many post-holocaust writings, especially in science fiction. Outside the realms of fiction, there is more evidence to support this view. In the prolonged controversies about the dangers of radioactive fallout of the 1940s and 1950s, two fearful possibilities loomed large. One was an increase in cancer after exposure to radiation, the other the prospect of mutations. Not only was fallout invisible, odourless and tasteless, but its effects might be felt by generations yet to come. Again, anxieties about the most apocalyptic of physical technologies often focused on possible biological effects. Later, concerns about nuclear power plants often took a similar shape.

The discussion of the potential consequences of nuclear fallout promoted public awareness of a number of the biological facts of life, as they were then understood. There arose a greater interest in genes and mutations, and a renewed emphasis on the centrality of genes in determining characteristics. Popular texts explaining the effects of fallout commonly included some elementary genetics.[10] The biologically neutral term 'mutation' – merely the motor for natural selection – had begun to acquire much more highly charged popular meanings before the war, but they were now greatly amplified. The grotesque deformities which resulted from mutation in a pre-war story like 'Master of the Genes' now became a cliché.[11]

In post-war science fiction, the associations of mutation were usually, but not always, negative. Stories about the end of all things may also be about new beginnings,[12] and mutation was sometimes used as a device for suggesting that while the Bomb would end an old civilisation, it would also usher in a new stage of human evolution; not just six toes, but telepathy and other marvellous powers would take humanity to a new level, transcending its origins in nuclear disaster. The point remains that it is a biological speculation which animates many of these stories.

These hopeful stories were the exception in printed fiction, and were mostly absent from 1950s cinema. The overall effect of the general outpouring of fear and guilt in the lightest of fictional disguises irrevocably skewed the meaning of mutation. It colours every lay reading of accounts of molecular biology, every genetic counsellor's consultation. Mutation means monstrosity. And any proposal for deliberate change in the human constitution evokes echoes from the time when it first seemed possible that the genome might be irrevocably altered by radiation.

I agree with Weart, then, that while Hiroshima is a cultural watershed, there are are also continuities in the influences shaping attitudes to science and technology – both physical and biological. But I disagree about which continuities are most significant. For the story I want to tell, what matters is that existing biological preoccupations were now put in a new context, one in which everyone's genes were at least potentially subject to radioactive bombardment. That, I suggest, is a part of the nuclear story. It is also an important part of the story of public debate about biology. We can trace some of these continuities more clearly by taking a closer look at the life of a biologist whose career and concerns embraced these issues before and after the war: Hermann Müller.

Saving our genes

It was Hermann Müller, the most prescient of geneticists active before the war, who did most to forge the link in public thought between radiation and mutation. It was Müller, the discoverer of the mutagenic effects of X-rays, who got people as worried about the genetic effects of nuclear weapons as about their explosive power.[13] Although Müller's remarkable career was rooted in the classical genetics of the 1920s, it was in the years immediately after the war that his public voice was heard most clearly. And the reasons he was so well fitted to be the propagandist of a new genetic decay were closely allied to the basis of his influence on his fellow biologists, an influence strongly felt in the genesis of molecular biology.[14]

Müller started out with a conviction that directed genetic change, leading to human control over evolution, would one day be possible. This was his motivation for joining T.H. Morgan in the 'fly room' at Columbia where the principles of classical genetics were being elaborated between 1905 and 1915. As he later recalled, he believed that 'this would provide a surer foundation and backing for a later attack on more specifically human problems. Only so could the necessary knowledge, as well as the authority, be obtained.'[15] Müller thus brought a vision of ultimate control over living systems into the heart of early genetics. As early as 1910, he formulated the rationally based eugenic programme which he was to restate at intervals for the next fifty years.[16]

Although hereditarian ideas were widespread at this time, perhaps the high point of the American eugenics movement, classical geneticists of Morgan's school generally distanced themselves from what they saw as the poor science of the human genetic studies conducted from establishments like Charles Davenport's Eugenics Record Office at Cold Spring Harbour.[17] The socialist Müller was additionally wary of the reactionary and racist cast of most eugenic thought, and rarely made his own views public early in his career. But he never lost faith in the possibility of a scientifically based eugenics. When the extremism of the leading American eugenists gradually led to their isolation from mainstream genetics, Müller played a leading role in the small group of liberal and radical geneticists – Kevles calls them reform eugenists – who attempted to reconstitute the scientific basis of eugenics in the late 1930s.

However, this went largely unnoticed by the wider world. Before World War II, Müller's energies were largely concentrated in the laboratory, with occasional non-scientific lectures and articles on social and political problems. His most famous discovery, the induction of mutations by X-rays, found its way into the science fiction magazines of the late 1920s, as we have seen, and made a considerable public impact. He published his first paper on the subject, with few details, in *Science* in July 1927.[18] And when he gave a fuller account to the Fifth International Congress of Genetics in Berlin later that summer, in an hour-long address on 'The problem of genetic modification', he created a sensation. According to his biographer,

> The press despatched the news around the world. Man's most precious sub-
> stance, the hereditary material . . . was now potentially in his control. X-
> rays could 'speed up evolution', if not in practice, at least in the headlines.
> Like the discoveries of Einstein and Rutherford, Müller's tampering with
> a fundamental aspect of nature provoked the public awe.[19]

From now on, Müller's eugenic concerns were mainly expressed through a preoccupation with the effects of radiation, a subject which he would make his own after World War II. But despite his momentary celebrity, he was not yet ready for public prominence. Indeed, in 1932 he suffered a nervous break-down, and on recovery moved overseas, first to Berlin, then Moscow. In Russia he had hoped for a more congenial political climate, but soon became embroiled in the controversies surrounding Lysenkoism until eventually he left and made his way back to the US via Spain and Edinburgh. This pro-longed absence from America helped keep him out of the public eye. His book *Out of the Night*, subtitled 'a biologist's view of the future', was published in the US in 1936 while he was in Moscow, but its combination of socialism, genetics and evolutionism was not to the public's taste.

His second major public statement before the war was at the Seventh International Genetics Congress in Edinburgh in 1939. Müller had been brought to Edinburgh by Julian Huxley as a refugee from the Spanish Civil War, and taught there for three years. At the Congress he drafted the response to a cabled query from the editor of *Science Service*, the American scientific news agency: 'How could world's population improve most effectively genetically?' The reply, signed by Huxley, Haldane, Lancelot Hogben and Joseph Needham, among others, was a 'Geneticist's Manifesto'. Viewing a world apparently on the brink of dissolution, they nevertheless offered a document reaffirming their faith in positive eugenics, while emphasising that 'the genetic improvement of mankind is dependent upon major changes in social conditions, and correlative changes in human attitudes'.[20]

The expected war broke out as the conference adjourned, and few knew or cared about the future genetic condition of humankind for the next six years. After the war, Müller, like so many other scientists, found his life transformed by the Bomb. But his preoccupations were still the same. In 1946 he was awarded the Nobel Prize for physiology or medicine for his pre-war work on mutation. His views now in great demand, he eschewed talk of positive eugenics for the time being, conscious that as the true extent of the Holocaust in Europe was revealed, popular revulsion precluded advocacy of any eugenic programme.

On the other hand, the widespread anxiety about the atomic bomb produced an audience eager for any authoritative information about its effects, and Müller quickly turned to providing it. His new series of warnings about the need to protect the human germ plasm from radiation began with his Nobel lecture in 1946,[21] and continued until he died in 1967. All along, he was as much concerned with routine exposures to X-rays, in medicine and dentistry or from the fluoroscopes then used in shoe shops, as with fallout hazards. But it was fallout which gripped the public, and he was ready to respond. Müller's articles, with titles like 'Time bombing our descendants' and 'Race poisoning by radiation' did much to publicise the connections between atomic energy and fundamental biological concerns.[22]

In addition, Müller used his new prestige and authority to exercise a less visible but no less important influence on public education, during his later membership of the Biological Sciences Curriculum Study (BSCS) programme. This was part of a major overhaul of American science teaching begun in the late 1950s, and speeded up after the Russian launch of Sputnik in 1957. According to Carlson, 'Muller attended all of [the BSCS] organization and planning meetings and read through all of the versions it produced, holding out strongly for and obtaining an emphasis on genetics and evolution.'[23] Müller's unshakeable conviction of the centrality of genetics thus helped to shape the educational programme which would prepare the next generation of Americans to grapple with the new discoveries of biology of the 1960s and 1970s.

Those discoveries turned out to be a realisation of Müller's earlier scientific vision, and his long-running influence on the direction of biology was now becoming apparent. His wish to uncover the mechanisms of inheritance led him to propose the elucidation of gene structure as the crucial problem for the future of biology; it made him one of the first disposed to think in the terms which would later come to define molecular biology. His frequent pre-war scientific addresses on themes like 'The gene as the basis of life' stimulated interest in the gene as a central organising principle of living things,[24] and his prescient observations about the need to 'grind genes in a mortar and cook them in a beaker' foreshadowed the route which others would follow toward a physico-chemical understanding of the genetic material.[25]

After the war, then, geneticists indeed became 'bacteriologists, physiological chemists and physicists', as Müller had foreseen in 1921.[26] His own role was now to promote a continued interest in the relation of fundamental genetic knowledge to human evolution. Thus, in 1949, when the most active work was on the genetics of bacteria and their viruses, at the far end of the scale of living things, Müller became founding president of the American Society of Human Genetics. He outlined a new research programme which included studies in human chromosome mapping, human genetic aberrations, and cytogenetic studies on human cell and tissue cultures.[27] These efforts helped to ensure that human genetics did not lag too far behind molecular biological work and that the new insights of bacterial genetics were eventually assimilated into theoretical studies of the human genome. This in turn helped to form the climate of speculation about possible practical applications of the new biology to human beings which emerged in the 1960s, and in the end to turn this speculation into the real work now under way in the Human Genome Project.

Müller's story was only one of countless others which were transformed by the advent and use of the atomic bomb. The coming of the nuclear age reaffirmed the power of the physical sciences, confirming predictions that had been made since the discovery of the nature of radioactive decay by Rutherford and Soddy in 1901.[28] But although the bomb's effect on attitudes to science and technology is hard to overestimate, the logic of Haldane, Huxley and their contemporaries was still good. To those who heeded their message, the advent of atomic energy, rather than diverting attention from the life sciences, merely offered a new comparison for their thinking about biology. As early as 1949 the prolific British science writer J.G. Crowther told the World Federation of Scientific Workers that 'the implications of atomic energy were impressive enough, but we should expect in the future biological discoveries of still greater implications'. These might extend to 'the possibility of directly changing living things, including man himself'. And as well as the warning of Hiroshima, there were now other examples of what the use of science might mean: 'When we remember what Nazi scientists have already done in their extermination camps,' Crowther continued, 'we would be utterly horrified by what would happen if these future discoveries fell into the wrong hands.'[29]

Both these rhetorical ploys, the analogy with nuclear energy and references to concentration camp experiments – notably by Mengele at Auschwitz – would reappear in later discussions of biology. They were, however, rarely heard in the years immediately after World War II, when biology was generally overshadowed by other sciences as far as the public was concerned. The preoccupation with atomic physics was evident in Aldous Huxley, for example, who understandably singled out atomic weapons as the most notable omission from *Brave New World,* in a new preface written in 1946. He now felt that nuclear energy was destined to disrupt all existing patterns of human life; 'Procrustes in modern dress, the nuclear scientist will prepare the bed on which mankind must lie'. Even so, in the same preface he reiterated that it was the sciences of life which had the greatest potential to revolutionise human life.

A partial exception to the concentration on physics was the attention devoted to the simple cybernetic devices of Ross Ashby and Grey Walter, and to the first predictions of imminent development of thinking machines. In this area, the situation after World War II bears comparison with that in biology before World War I: a few workers became highly visible because they made good newspaper copy. They had simple devices to demonstrate, and large claims to hang on them. The work was easy to communicate, and commentators were ready with far-reaching predictions. The long-standing public fascination with machines which appeared to imitate life was reinforced.

Rosenberg gives examples of the reporting of these electromechanical 'animals' which are in the same vein as earlier reports of the 'creation of life' or the 'glass heart.'[30] Thus Ross Ashby's homeostat was described in *Time* magazine in 1949 as 'The Thinking Machine'. Even more remarkable abilities were attributed to W. Grey Walter's series of mechanical 'tortoises', which *Time* depicted as roving about Walter's house, seeking food to meet their photo-electric hunger, and as showing symptoms of neurosis when given contradictory stimuli.[31] The idea of the thinking machine was expanded in a book by E.C. Berkeley called *Giant Brains or Machines that Think*, part of which appeared in the newly launched *Science Digest* in the USA.[32] There were also some hostile reactions to these predictions of imminent scientific achievement.[33]

Both nuclear technology and artificial intelligence would be discussed many times over the next half-century, but very often in terms that derived from the earlier history of debates about biology. The common elements which tend to appear in relation to all these fields mainly stem from images of biology, with Frankenstein the most widely applicable motif. Mythical stories or scripts of this kind can be remarkably fuzzy. They overlap, and cross-fertilise. The effect of the fallout controversies on ideas of mutation is a clear-cut case, but there are others. Reports of Ashby and Grey Walter's devices, for example, which show the enduring interest in mechanical creatures, call to mind Jacques Loeb's fascination with the selenium-eyed 'dog' built by J.H. Hammond in the 1900s.[34]

Meanwhile, biology proper occasioned little comment. There were brief press flurries about one or two of the early hints that the borderline between life and non-life was coming under investigation, like Miller's work on the chemical origins of life and Fraenkel-Conrat's reconstitution of tobacco mosaic virus in the test-tube from dissociated RNA and protein.[35] But neither endured. Nor were there many fictional efforts to explore biological ideas, aside from the armies of mutants. Even John Christopher's *Death of Grass*, in which a mutated virus attacks the world's grain harvest, is a post-nuclear novel in disguise – although it does anticipate some of the disaster scenarios which were put forward during the recombinant DNA debate twenty years later.[36] Apart from these, *Brave New World* and, for that matter, *Frankenstein* and *Moreau* were still being read, and plagiarised. But there was no real advance in fictional ideas to compare with Huxley's book.[37]

Nevertheless, the laboratory work that would lead to the new approach to biology which we now know so well was maturing rapidly. It was not yet visible outside the scientific journals, but in just two decades' time, in 1968, the geneticist Richard Lewontin wrote with confidence that

> there is not the slightest question that more progress has been made in our understanding of the fundamental physico-chemical mechanisms of life in the last twenty years than in all the previous history of biology. Nor is there any doubt that the movement that calls itself 'molecular biology' is in large part responsible for this progress.[38]

By the late 1960s, also, the advent of molecular genetics had been incorporated in a public notion of a wider 'biological revolution', portending in fact what Huxley had envisaged in fiction. The next chapter describes how that idea took hold.

Chapter Seven

Priming the Biological Time Bomb

> Many of the revolutionaries are caught up in a fervor not
> normally seen in scientists, who ordinarily speak cautiously.
> Columbus wasn't cautious and in substantial numbers they are
> not. They are now certain that new worlds can be discovered
> inside the human being. And they are eager to get on with the
> discovering. They see themselves out on the cutting edge of
> science.
>
> Vance Packard, *The People Shapers*, 1978

It is one measure of the importance we have been taught to ascribe to the rise
of molecular biology that it is now widely assumed that its earliest successes
were well advertised. Later journalistic mythology has it, for example, that
the discovery of the DNA double-helix in 1953 created a major stir right away.[1]
Not so. The first article in the press to describe the double helix, by Ritchie
Calder, appeared on 12 May 1953, several weeks after Watson and Crick's paper
in *Nature*. And as Edward Yoxen remarks, it was portrayed as 'an isolated
event by anonymous scientists in an unidentified laboratory. Molecular
biology as a science with a publicly recognized connotation [was] not yet in
sight.'[2]

At first, the new work, in all its abstraction, appeared remote from everyday concerns. As with Wendell Stanley's crystallisation of tobacco mosaic virus
before the war, the complexities of bacteriophage genetics and problems of
crystallographic analysis of macromolecules were not of any obvious relevance
to the power of biology to affect human lives. But over the next decades, we
were all brought to see life in new ways. The idea that molecular biology had

uncovered the 'secret of life' was promoted assiduously by some scientists and popular writers. The question is how did the reading of these secrets relate to earlier ideas about science and life?[3]

Part of the answer lies in the motives of the pioneers of molecular biology, which were in many ways continuous with the style of work personified by figures like Jacques Loeb. In some respects, the abstruseness of biological work after the war, the concentration on ever smaller elements in the organisation of living things, were products of new kinds of questions. But they were still driven by older agendas. As the distinctive concerns of molecular biology began to be defined, they became the shared problems of a group of workers who were interested in exploring 'the borderline between the living and the dead', in Francis Crick's phrase,[4] guided by fundamental principles of physics and chemistry. These researchers, in Britain, France and the US, came to formulate their questions in terms of the relation between structure and function at the molecular level. And they were quickly rewarded with startling success in understanding the chemical structure of the gene, and its role in transfer of information in the cell. At the same time, a new set of ways of investigating the operation and regulation of all cellular processes – metabolic, immunologic or genetic – in terms of macromolecular specificity – were developed.

The temptation to see the moment of the double helix as a complete revelation, immediately broadcast to the global media, to divide biology into pre- and post-double-helix eras, stems in part from the great simplicity and beauty of the conceptual scheme which dominated molecular biology by the end of the 1960s, and which still largely dominates today. The unity of structure and function of the genetic material is easy to grasp. The biochemical and structural features of the double helix immediately suggested a mechanism for the transmission of genetic information from generation to generation, and provided the basis for investigation of the flow of information within the cell: the way in which the gene specifies the structure of the appropriate protein. But while the full working out of what Keller calls the discourse of gene action offered a captivatingly clear picture of the mechanics of gene transcription, translation and replication,[5] the path from the double helix to this complete paradigm was complex and sometimes tortuous. In particular, after the moment of clarity when the DNA structure was first proposed, a host of problems which could only now be formulated preoccupied the new biologists for the rest of the 1950s. How might a genetic code be written and read? How could gene action be regulated? What did mutations do to gene structure?

Teasing out the answers depended on sophisticated theoretical arguments and delicate experimental work on the genetics of *E. coli* and its bacteriophages. Little of this work was accessible to the public, partly because the state of the science remained so fluid that few outsiders attempted to keep track of events.[6]

A good example of the difficulty of conveying the exact state of current work is Beck's *Modern Science and the Nature of Life* of 1957, one of the first popular works to take account of the new approach. The book is an erudite and witty overview of the history of biology, seen as a record of progressive enlightenment as rational inquiry gradually infiltrated biological thought. Towards the end, in a chapter headed 'The living organism today', Beck describes some of the origins of molecular biology, emphasising the speed and scale of the transformation of understanding that was now under way.

He begins by contrasting the current state of the art with the outline of biology presented a generation earlier in Wells and Huxley's *The Science of Life*. This he considered a reasonable statement of the biologist's frame of mind in 1929. But in all the wealth of material in those three volumes, 'what is missing . . . is a sense of structure and direction, a clearly defined centre of gravity around which all other ideas may be balanced'.[7] Now, however, 'the intervening years have clearly revealed that there is one question in biology that transcends all others – a question that science needed insight to ask and that it may soon be ready to answer. It is the question of the gene.'[8]

He goes on to explain how the recognition of 'the question of the gene' as a key problem came about and discusses its biochemical structure. But the large claim that the problem is about to be solved is backed up by a burst of waffle; no specifics. In 1956, all Beck could say was that 'we cannot . . . enter into the details of this work . . . since these are questions that science is investigating at the hour of this writing . . . Today, 1950 seems like the middle ages, for the situation is moving more rapidly than ever. Tantalizing developments are literally tumbling from the laboratories.'[9]

So Beck was moved to offer one of the first loud advertisements that a new biology was on the way, but could convey little beyond a general sense of excitement. The field was moving too rapidly to pin down. Full elucidation of these 'tantalizing' developments would soon be offered, but the main outlines of a new way of describing life had yet to become clear enough to relate for the general reader.

A familiar question posed anew – can man be modified?

Beck's vision of the imminence of a new biological era came at the height of the Cold War, a time when the idea that science had ushered in the nuclear age was still a dominant theme in popular commentary. In retrospect, though, the primacy of the Bomb as the most imposing symbol of the power of science was relatively short lived. The unleashing of a force which could bring about universal destruction, often foreshadowed in fiction, was the apotheosis of physical scientists' technological achievement. Now that it had been realised, new physics or chemistry would never be as impressive again. Awesome and terrible as the nuclear arsenals were (and are), the destruction they might bring was in one sense the ultimate achievement in conscious mastery of the environment; they could render it uninhabitable. But if the Bomb was not used, life would go on, even if under a nuclear shadow, and the logic first spelled out clearly by Haldane and Huxley gradually reasserted itself: the science with the greatest potential for altering human life, as long as it might continue, was biology. As time passed, the existence of nuclear weapons came to be incorporated into the vocabulary used for discussing biology, as in the early example cited by J.G. Crowther quoted on p. 132. The observation that the physical sciences can only change the external circumstances of human life, but the sciences of life may transform humanity itself, became more widely heard. Yet the physical sciences had already furnished the means for self-destruction. What, then, could be expected of biology in future? This was the question which the generation of popularisers who followed Beck, the prophets of a new biological revolution, tried to answer.

Towards the end of the 1950s, there appeared evidence of a wider perception that biology had made progress since the war. Popularisation of the life sciences was already benefiting from a heightened interest in science stimulated by the outcomes of the physicists' war in nuclear weapons and the controversies over fallout. New publications like the Penguin *New Biology* and, later, *Science News* in Britain and *Science Digest* in the US were promoting a wider appreciation of how far understanding had already advanced in such areas as cell structure, endocrinology and animal behaviour. At the same time, as the problems posed by the Bomb became, as it were, part of everyday life – at whatever psychological cost – there were signs of a renewed concern with the problems that might be presented by future biology. The *Bulletin of the Atomic Scientists*, for example, began to publish occasional articles on biology. One of the earliest, on 'The warning and the promise of experimental embryology',[10] came well before the *Bulletin* began to publish pieces on molecular biology.

This reawakening of concern may be viewed through an examination of one particularly influential text. In 1959 there appeared an English translation of a book written three years earlier by the French biologist Jean Rostand, *Can Man Be Modified?* This book and its reception both point towards the resurgence of discussion of the social impact of biology which was to take place in the 1960s. Rostand had followed the course of earlier debates about biological control and was familiar with up-to-date laboratory discoveries, but his book was largely a compilation of examples and episodes I have already discussed. It took no account of the new molecular biology, and made no mention of the identification of DNA as the genetic material. The text thus offers a review of some past landmarks in the public history of biology, and helps us to assess the state of public debate about the subject before the explosion of comments on molecular biology in the following decade.

Rostand was well fitted to give an overview of the century's biology. Born in Paris in 1894, the son of the celebrated poet and playwright Edmond Rostand, he became a pioneer experimental geneticist. As a young man he read Bernard, Pasteur and Yves Delage and, by his own account, was led to specialise in the study of frogs and other amphibia by Bataillon's work on artificial parthenogenesis in 1910. From this point on, as he put it, 'the whole of experimental biology, practically speaking, [was] built up under my eyes.'[11] Aside from work on the genetics and reproductive biology of amphibia, he also contributed to human genetics, and co-produced an atlas of human genetics.

The atlas was only one of more than fifty books from Rostand's pen, mainly on biological and philosophical topics. Outside the laboratory, he applied his considerable literary gifts to become the best-known biologist and populariser of his time in France. Among his other biological works were the pre-war *Adventures before Birth* (1936), which was translated into English by Joseph Needham, and *Life, the Great Adventure*, published in English in 1955, which presented expanded versions of a series of radio talks.[12] He thus emulated Huxley in his mix of media.

Can Man Be Modified? was a slim volume, just over 100 pages, in three loosely connected sections. In the first, 'Victories and Hopes of Biology',

Rostand provided a quick review of relevant research. His first example was artificial parthenogenesis, citing Loeb and Bataillon. This led on to the possibility of novel modes of reproduction, quickly brought up to date with the nuclear transplantation experiments of Briggs and King in 1955. This was one of the first public accounts of this work, the foundation of modern research on asexual reproduction by cloning, and already Rostand felt justified in concluding that 'this new technique of generation would in theory enable us to create as many identical individuals as might be desired. A living creature would be printed in hundreds, in thousands of copies, *all of them real twins.* This would, in short, be *human propagation by cuttings . . .*'[13]

The exposition continued with other recent research: work on preservation of mammalian semen at low temperatures and the possibility of artificial insemination using frozen semen from a father now dead; induction of sex changes by hormone treatments and other possibilities of hormonal control over development; improvements in techniques of organ culture, described as 'begun in 1936 by Carrel and Lindbergh', which enabled Wolff to culture embryonic organs of chickens and mice. This last, Rostand suggested, might 'contain the rudiments of ectogenesis, or test tube pregnancy, which Aldous Huxley promised us in *Brave New World.*'[14] Finally, he mentioned production of artificial mutations, a way of causing 'new races of living creatures to appear on the earth'. This selection of findings emphasised the growth in biology's manipulative power. As Rostand said, 'from now on, we possess the means of acting on life'. True, he also stressed the limits of present-day knowledge, quoting approvingly from Bataillon that 'when we succeed it is because, on some imperceptibly small point, our logic is in conformity with a logic that goes prodigiously beyond us'. But the implications of this sudden access of modesty were ambiguous. On the one hand, he suggested that we should not set too much store on past achievements when the unsolved problems remaining were so large. On the other, if a biology only partially developed is as impressive as he describes, then the prospect for control over life in future must be more impressive still.

Like Beck's more serious history, Rostand's little book stood at the transition between the self-conscious humility in the face of the complexities of living things which predominated before the war and the far-reaching self-confidence which marked the expansion of molecular biology. Rostand stopped short of asserting that the possibilities for greater control would certainly be realised, but as an experimentalist he looked forward eagerly to the attempt:

For the moment, biology is incapable of satisfying the principal requests addressed to it. It does not prolong life, it does not determine sex, it does not control heredity, it does not procure intelligence for fools . . . But it is possible that all these powers may belong to it tomorrow, and many others as well . . . the procreation of twins at will, test-tube pregnancy, the modification of the embryo, controlled mutations, the production of a superhuman being.[15]

The last three items on this list look forward to the third and final part of the book, in which the question posed in the title was discussed in more detail. First, however, Rostand offered a more general reflection on man and science. Here Rostand the philosopher ruminated on the uses of scientific knowledge, devoting as much space to computers and cybernetics (including Grey Walter's tortoises), and the global implications of atomic energy as to biology. In the end, though, it was biology which was the most significant, and for the reason we now expect, a reason Rostand puts in Carrel's words from *Man the Unknown*: 'humanity has become the master of its own destiny . . . it is both the marble and the sculptor.'[16]

But Rostand takes the argument a step further than his compatriot. We must wonder, he suggests, 'whether science is not on the point of reaching a kind of frontier, beyond which its progress might prove more harmful than advantageous . . . Perhaps human life should go on propagating itself in the shadow and science should never throw on it the beams of its intense light?'[17] Rather than answer the question directly, he retreats into generalities, observing that 'to attenuate the dangers which are the price of our conquests, we shall need all the conscience that is available'. It is a position which prefigures that taken by many writers on the Human Genome Project thirty-five years, but a whole biological era, later on.

The third part of *Can Man Be Modified?* is essentially a recapitulation, now focused more directly on the possibility raised by the title. Once again, the uses of chemicals, hormones, psychosurgery and ectogenesis are canvassed. And again, *Brave New World* is used to set the scene, with long quotations from the opening description of the hatchery. Rostand recalls Needham's review of the novel, asserting that 'this vision of the future is based on a precise knowledge of the present', and once more cites Carrel and Lindbergh in support.

He then turns to a review of eugenic proposals, including Müller's, to which he gives his qualified support. This kind of programme is seen as merely transitional, however, leading towards a future when 'we shall learn how to make human hereditary characteristics mutate in a direction that will be both predictable and advantageous. When that day comes man will be able to modify himself when and how he wishes.'[18] From here, the chapter moves on to a rather abstract discussion of the morality of trying to create a superman, one aspect of a wider question: 'that of man laying hands upon himself, of the application of biological techniques to the human person'.

The brevity of the book and its repetition suggest that speculations about manipulation of human beings still rested on a fairly shallow scientific base. Rostand reviewed a range of past and present developments, but the complete catalogue is fairly small. However, he pulled together concerns which had been developing through the century, and provided a convenient source for many later writers, who would have far more scientific evidence to draw on. The way in which quotations and anecdotes from *Can Man Be Modified?* recur throughout many later, longer books on the same theme identifies Rostand's inquiry as a link between the time of Loeb and Carrel and the biological revolution of the 1960s.

As with *Brave New World*, reviews of the English translation of Rostand's book suggest a more widespread concern about the possibilities of biological control. It was praised by the *New York Times* for an 'unusual readiness to speculate'.[19] Others took Rostand's claims more seriously. The *Times Literary Supplement*'s reviewer commented that the book showed 'how near we have already come to the *Brave New World* of Mr. Aldous Huxley'. The reviewer also observed that Rostand's idea that 'homo sapiens is in process of becoming homo biologicus' was 'very exciting as well as a little disturbing', and that the possibility of creating a superman was a 'bewildering prospect'. The final comment cast doubt on Rostand's optimism that the moral problems thus posed were soluble.

> M. Rostand tries to calm the reader at the end by assuring him that the future is guaranteed by love, which science recognizes as an essential attribute of the species and the necessary foundation for science itself. It would be nice to think so.[20]

Similar doubts were evident in the *New Yorker*, which suggested that 'the ultimate questions about values and about the relation of scientific progress to the good life remain unanswered, and usually unasked', in what the reviewer found a 'brilliant and deeply disturbing book'.[21]

The disturbance was shared by the zoologist Anthony Barnett, in the *New Statesman*. He was happy to see Rostand's question raised, but felt that the book was weak on substantive comment. In his view, the science had moved further than the book revealed and 'the reader will not learn just how far embryologists, neurologists and others have gone towards founding an actual *Brave New World*'. This was unfortunate, because 'M. Rostand has some good points to make and a subject – the biological control of man by man – which urgently needs public discussion'.[22] When that discussion came, though, writers warning of the potential problems raised by biology would generally receive a more critical reception. For now, no one was claiming that such visions were evidence of an 'anti-science' movement. That line of argument would only surface when research appeared to be under threat, a development which followed the full realisation of the promise of molecular biology.

The secrets of life

In the decade after Rostand's book became available in English, the public face of biology was transformed. Both the significance ascribed to the science and the language in which it was described were permanently changed. Part of the change resulted from the advent of molecular biology as the dominant way of seeing living things, but advances in more traditional disciplines also contributed to the mass media depiction of a new era in biological technique. The 1960s was the decade of the Biological Revolution.

Perhaps the most general condition needed for this idea to take hold was a recovery of scientific optimism among biologists. The self-confidence of its practitioners has always been a hallmark of molecular biology, right from the earliest days of the phage group, most of them originally trained as physicists, just after the war. This group, who agreed to study the bacterial viruses known as phages, were impatient with traditional approaches to heredity and convinced that a reappraisal of the problems in physico-chemical terms was needed. After the formulation of the DNA structure in 1953, this confidence in the prospects for improving fundamental understanding of living things came to be shared by many young biologists who believed that the tools now existed for a concerted attack on the key problems of their science. There was, in fact, a recovery of the optimistic spirit of Jacques Loeb's day, after the modesty which seems more characteristic of the biologists of the 1920s and 1930s. There was also a resurgence of emphasis on the kind of mechanistic explanation advocated by Loeb and some of his contemporaries. This time, the mechanistic explanations were to be put forward and tested at the molecular level.

This time, too, the confidence was largely justified. There followed what was, by any standard, a brilliantly productive phase in biological research. As Horace Judson, the indefatigable chronicler of these times, summarises:

> By 1970, a coherent outline of the processes of life had been put together: what the gene was, how genes duplicated and mutated and how they were expressed in the assembly of proteins, how the expression was controlled, how proteins functioned and interacted – how the organism built itself and perpetuated its kind.[23]

By 1970, this story was also the common currency of journalism and popular writing. Judson's massive oral history is itself a popularisation of the molecular biologists' world-view, but he was a relative latecomer, a journalist drawn to the new developments in the 1970s who took the job of describing them further than anyone else and turned into a historian in the process. But the confidence of the new-style biology began to become evident to interested publics in the 1960s. As Edward Yoxen puts it:

> There came into being in the late 1950s and 1960s a new way of representing biology. One has only to collate the books, articles, discussions, conferences, films and radio programmes made about the 'New Biology' to see the continued recurrence of phrases like 'cracking the code of life', 'the Biological Revolution', and the 'New Frontier in biology', which imply not merely a greater understanding of the processes of life, but also a profound transformation in the relation of man and nature and the beginning of a new era in human history.[24]

As this suggests, both new and old media were important in bringing the new biology to wider notice. The Sputnik launch in 1957 helped generate interest in a number of new ventures in science reporting, particularly on television, and these soon began to take note of the developments in molecular genetics. In Britain, two major programmes discussing the new work in biology appeared on BBC television before the end of the 1950s: *The Thread of Life: DNA* in 1958 and the wider-ranging *What is Life?* in 1959. The latter was the first science programme in the UK to attract more than 10 million viewers.

But at this time, as Beck's book showed, the subject was still in flux. So *What is Life?*, for example, did not bring out fully all of the implications of the new approaches in biology. Although the implications of the possible synthesis of life were aired at the end, the programme was mainly devoted to illustrating the idea that a radical transformation was taking place in the scientific conception of life, without trying to assess its significance.[25]

A few years later, the mass of new information had been put into a more orderly framework. In 1957, Francis Crick asserted the primacy of two simplifying assumptions, even though 'the direct evidence for both of them is negligible'.[26] Five years later, both of these principles – the sequence hypothesis (that the linear order of bases in a stretch of DNA encodes the amino-acid sequence of a particular protein) and the so-called central dogma (that information can only pass from DNA to protein, not vice versa) – were held to be proven. Furthermore, the solution to the genetic code was in sight, what Judson calls the 'end of the beginning of molecular biology'.

For the wider world, there was also the fivefold Nobel Prize award of 1962, when Watson, Crick and Wilkins received the prize in physiology or medicine for the DNA structure and Perutz and Kendrew the chemistry prize for their work on the haem proteins. This helped crystallise the shift in the focus of public awareness of biology. And the award in physiology or medicine already seemed to promise some practical application for these seemingly esoteric studies.

As in previous periods when professional and public ideas about experimental biology combined to generate high expectations, attention came to focus on a few practical goals which came to symbolise the potential manipulative power of the science, for both biologists and laypeople. The two main goals were not new, but were now conceived in new ways. The ancient idea of creating life was now to be equated with artificial synthesis of DNA, displacing protoplasm as the essential ground substance of living things with a rather simpler chemical. Similarly, long-standing speculation about transforming species or modifying man came to centre on the possibility of manipulating the hereditary substance – soon to be christened 'genetic engineering'. These three related ideas – the centrality of DNA, the notion that life might be created by synthesising DNA and its forms altered by altering DNA – were major motifs in the discursive literature about molecular biology which grew up from the beginning of the 1960s. The first contributions were largely the work of scientists, but their ideas were quickly taken up by other writers. Throughout the decade, journalists and popular authors responded rapidly to the latest biological developments, and translated them into the terms of these ideas. Biology was once again news.

As the decade advanced, the volume of such writing expanded enormously, so a few examples must stand for many. The early translation of scientific optimism into popular terms is well shown by a front-page story in the *New York Times* in 1962, headed 'Biologists hopeful of solving secrets of heredity this year'.[27] Asserting that 'the breaking of the genetic code will rank among the greatest scientific achievements of all time', the article discussed future prospects thus opened up. These included 'controlling the inheritance and hence the destiny of plants and animals and perhaps even man himself', and 'creating life, of a sort, from chemicals that today are in bottles on the scientists' laboratory shelves'.

This second possibility was discussed at the time of the Nobel awards later that same year. The *Sunday Times Magazine*, in a piece on 'The architects of life', suggested that 'their work heralds the day when man may be able to create simple forms of life himself at will, in the test-tube'.[28] These predictions were repeatedly endorsed by scientists, as in a feature in the *New York Times* in December 1962 by staff writer Lawrence Galton. He interviewed a number of researchers at Columbia University about the prospects for the next thirty years. Both the biologists he spoke to, neither of them major figures, were confident that great successes lay ahead. Dr W. Sebrell felt that 'we're going to see within the next generation the artificial creation of living things starting with the virus – and that's almost here'. Professor T. Hayashi looked forward to a day when 'with refined instrumentation it may be possible to destroy a bad gene in a sperm head or take control of heredity by adding chemicals to the DNA molecule'.[29]

This passage sounds more like a journalist writing than a scientist talking, but similar prognoses were certainly offered by many other, more senior scientific figures. Hermann Müller, for example, in the *Bulletin of the Atomic Scientists*, compared the position of biologists in the early 1960s directly with the state of the science as envisioned by Jacques Loeb sixty years before. Loeb's two greatest aspirations, he wrote, had been to produce controlled abiogenesis and to effect the transformation of species, and 'within recent years, the bare essentials, at least, of both these objectives have been attained'.[30]

The conviction that molecular biology was yielding so deep an understanding of living systems that these manipulative goals were now attainable was heard often throughout the 1960s, and appeared to grow in strength. As ever, some scientists expressed great enthusiasm for the medical benefits which would ensue. Other writers, scientists and lay commentators, were more dubious. Many were concerned with the social, political or ethical problems to which such control might lead. The debates which would later attend the Human Genome Project began to appear clearly in outline at this time. This can be seen readily in the pages of the *New York Times*, the paper which had the strongest coverage of scientific affairs. The *Times' Index* between 1963 and 1968 records the establishment of this debate in reports and editorials about the future of biology. An editorial, prompted by the Eleventh International Congress of Genetics in 1963 in Holland, encapsulates in a few hundred words many of the concerns of a multitude of later articles. It suggests that the conference 'was probably the most closely watched scientific meeting of recent years'.[31] This the author attributes to the widespread belief that 'in the next few years humanity will understand – and be able to control at least in part – the fabulously intricate mechanisms through which each species of living organism transmits its essential properties to the next generation'. When this genetic breakthrough occurs,

It promises to pose problems man has never faced before – problems more difficult in some ways than the gigantic ones posed by the mastery of nuclear energy and the beginnings of space flight. Ultimately, we may be able to fashion living species to order, to 'manufacture' living organisms with specific properties just as we now produce machines or instruments.

We must discuss what we do with these abilities, asking,

Is mankind ready for such powers? The moral, economic and political implications of these possibilities are staggering, yet they have as yet received little organised public consideration. The danger exists that scientists will make at least some of these God-like powers available to us in the next few years, well before society – on present evidence – is likely to be even remotely prepared for the ethical and other dilemmas with which we shall be faced.[32]

The article is typical in several respects. Many comments on the new genetics compared its implications with those of nuclear physics. But although there was strong concern, it was expressed generally and rather abstractly. New problems of various kinds are mentioned, but no attempt is made to specify exactly what they are. Similarly, there is a call for public discussion but no indication of what should be discussed, or what form the discussion might take. Expressions of such generalised concern tend toward repetition. The needle stuck in this groove through many similar articles, which concluded with almost identical pleas for more debate so that society would somehow be 'prepared' for the advent of genetic manipulation. Later statements quoted in the *New York Times* which follow this pattern include one from Dr F. Seitz, president of the National Academy of Sciences, from Philip Abelson, in an editorial in *Science* in which he quoted Aldous Huxley, from the geneticist Bentley Glass, and from Marshall Nirenberg, again from *Science*, in an editorial in which he predicted that animal cells would be programmed with synthetic messages within twenty-five years.[33]

At the same time, there was continued support from other eminent scientists for the idea of using genetic manipulation to cure disease , notably from Joshua Lederberg and Robert Sinsheimer, both of whom were often reported in this vein.[34] And a continuing succession of reports of actual laboratory achievements or proposed projects served to dramatise the growing manipulative power of biological science. Two of these drew especially wide attention. In 1965 Dr Charles Price, newly elected president of the American Chemical Society, proposed a massive national effort toward the creation of life, to rival the space programme. Two years later, Arthur Kornberg's group at Stanford succeeded in synthesising the DNA strand of a virus.[35] The latter illustrates well the changing constellation of causes behind claims made by or on behalf of scientists. The assertion that certain experiments represented steps toward the creation of life was no longer made to justify a philosophical stance, as in Loeb's day. Now, it was used to gain publicity for a university department worried about cuts in research funds. Yoxen concludes that 'the issue of creating life was deliberately included in a prominent place' in Stanford's press release.[36]

Whatever the shift in underlying reasons, the cumulative effect of these reports recalls that seen sixty years earlier. The constant reiteration of the power of the new techniques of biological investigation generated high expectations, allied to a generalised sense of concern or unease. The message was reinforced in books, television programmes and magazine articles. A large number of popular accounts of molecular biology appeared after the first outlines of the genetic code were clear. Those by scientists were more likely to stick to established facts, as in Kendrew's *The Thread of Life*,[37] while accounts by journalists were sometimes more freely speculative, but readily found scientists to draw on.

A good example of the popular writers' view is Lessing's 1967 book, *DNA: At the Core of Life Itself.* The book was based on a series of award-winning articles for *Fortune* magazine, and is thus more expansive than the newspaper items I have cited. But its basic outlook is the same. Lessing starts by asserting that 'Within the next few years, man is likely to take the first epic steps toward modifying directly his own hereditary structure', and goes on to discuss the immense powers for good and evil which this will unleash.[38]

He then runs through the by now fairly standardised story of molecular genetics with repeated emphasis on the manipulative possibilities it opens up. It is 'no longer difficult to imagine the day when chromosome analysis will be a routine part of prenatal care, and DNA abnormalities will be corrected in the foetus before birth', a possibility which 'raises large ethical, legal and social issues'. These are not, however, elaborated.[39]

Once again, scientists' own speculations are cited to reinforce the authority of the tale. During a later discussion of future possibilities Lessing quotes James Bonner of CalTech, known for his work on the function of the histone proteins in gene regulation.

'I have tried to think about what further organs I would like to have,' says Bonner only half jokingly, 'and I have decided that I would like to have four hands since there is so much for biologists to do. Recently, as I was trying to light my pipe in the laboratory, my colleague, Professor Huang, said to me, "If you're going to smoke a pipe in the laboratory you'll need five".'[40]

The impression given is that these men, who are actually doing the research which is transforming biology, quite normally indulge in such bizarre speculation. The disjunction between the everyday business of lighting one's pipe and the extraordinary notion of choosing how many extra limbs to order points up the sweeping claims being made, and the confidence of the scientists making them.

Awareness of the possibility of genetic engineering increased more or less in tandem with the spread of knowledge about molecular biology. And the idea of genetic manipulation gradually displaced the creation of life as the most common shorthand for the goals experimental biologists were now pursuing. Towards the end of the 1960s, the term 'genetic engineering' began to enter wider use, and to appear in more general contexts as a symbol of the powers obtained through scientific knowledge. In his controversial

BBC radio Reith lectures in 1967, the anthropologist Edmund Leach used genetic engineering as one of a series of motifs signifying the growing ability of science 'to tamper with nature itself – consciously and systematically'.[41] Genetic engineering and other forms of biological manipulation also began to appear more frequently in science fiction, as in Frank Herbert's 1966 novel *The Eyes of Heisenberg* in which an elite of immortals maintains control over the mass of the population by altering the genes of their embryonic subjects.[42]

By the end of the decade the idea of genetic engineering was probably as widespread as knowledge of the biological role of DNA. A survey published some years later, in 1976, gives an indication of the diffusion of these ideas. Winstanley conducted a small survey among British university students 'to get some idea of the extent to which DNA has been absorbed into the modern consciousness'.[43] She found that more than three-quarters of students had heard of DNA, and well over half could choose the correct definition from a small set of alternatives. Sixty per cent had heard of genetic engineering, a much higher percentage than had heard of Watson, Crick or the double helix. These were presumably a well-educated group, although the majority cited the mass media as their source of scientific information, and those who had studied biology in school did no better than the rest. The survey excluded those studying biology at university, who would presumably have raised the overall figures.

These findings, along with the popular texts of the time, establish two things. First, the new complex of scientific findings of the 1950s and early 1960s which made up molecular biology was communicated to lay publics pretty rapidly – if not quite the day after the double helix was discovered. Growing post-war mass media like television and illustrated colour magazines were well suited to popularisation of science, and even though molecular biology was visually unspectacular the pre-eminent post-war science was widely reported.

Secondly, this led not simply to a quantitative change in public awareness of experimental biology but also, to some extent, to a qualitative change, an alteration, perhaps, in popular consciousness of biological advances. This was in part a reflection of the attitude of the molecular biologists to their objects of study. But their message was amplified by journalists who appeared happy to take up the theme that 'life in its essence' was now understood. This understanding of living matter was to furnish a basis for new technologies, and perhaps for the engineering of new life forms.

1. The first image of Dr Frankenstein and his creature, from the 1831 edition
of the novel. Note the unmonstrous appearance, apart from his size, and the
scatter of bones in the high-windowed laboratory.

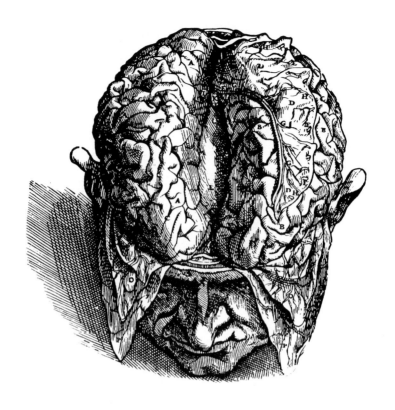

1. Eighteenth-century science made part of its public impact through spectacular lecture demonstrations, most famously depicted in Joseph Wright's characteristically-lit view of an experimental philosopher displaying the powers of the air pump.

facing page

2. *(top)* By the seventeenth century, educated men and women were familiar with the appearance of the body's interior, and readers of Frankenstein would have known the importance of anatomy for medical knowledge. This is Vesalius's demonstration of the first stage of dissecting the brain.

3. An engraving from 1818 showing electrical stimulation of an executed criminal's corpse – the kind of sensational demonstration which Mary Shelley certainly read about and possibly witnessed.

The EDISON
KINETOGRAM

VOL. 1   LONDON, APRIL 15, 1910   No. 1

SCENE FROM
FRANKENSTEIN
FILM No. 6604

5. The first screen creature: Charles Ogle as the monster in Edison's 1910
silent film.

facing page

6. *(top)* Boris Karloff's creature in Universal's 1931 movie had – and has –
extraordinary screen presence, both as a threat and, as here, through the
pathos of his mute appeals to his creator.

7. The *Illustrated London News'* depiction of Richard and Barnabas
Brough's *Frankenstein*, or *The Model Man*, a burlesque version of the
story which played at London's Adelphi Theatre from December 1849
into 1850.

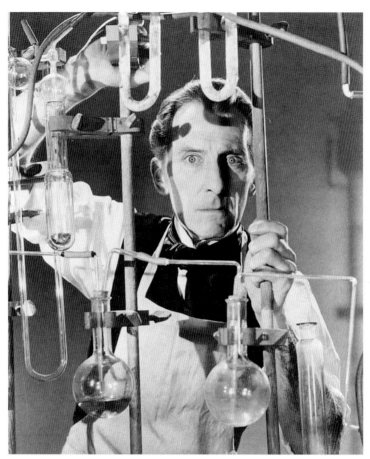

8. In Hammer's film series, the scientist, now Baron Frankenstein, was a more compelling character than his creation: driven, ruthless, brilliant. Peter Cushing is shown here in *Frankenstein Must Be Destroyed*, 1969, the fifth of seven pictures in the Hammer sequence.

9. The most articulate monster yet: Robert de Niro in Kenneth Branagh's *Mary Shelley's Frankenstein*, 1994.

12. Edward Schafer regarding his handiwork, the week after his Presidential Address to the British Association in 1912.

facing page

10. *(top)* Vivisector confronts victim: note the anaesthetic ready to be administered to the trusting dog, and the dead bird on the experimenter's table.

11. Experimental physiology and vivisection were inseparable, as here in Claude Bernard's laboratory. For Bernard, as for Dr Moreau, the animal was merely the structure which concealed the answer to his current problem.

13. The stereotypical biologist depicted in the frontispiece of Odham's *The Miracle of Life* in 1933. No creature is visible, only a man in a white coat with binocular microscope, dissecting needles, and chemical apparatus, searching for 'the secret of life'.

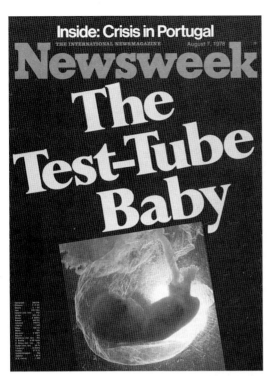

14 (a) *Newsweek* and (b) *Time* greet the birth of Louise Brown – the first 'test-tube' baby in 1978. One figures the event as a creation scene; the other uses the then novel photographic image of the foetus in the womb, a symbol of technology's power to probe reproduction and a near copy of the idea of the baby in the bottle.

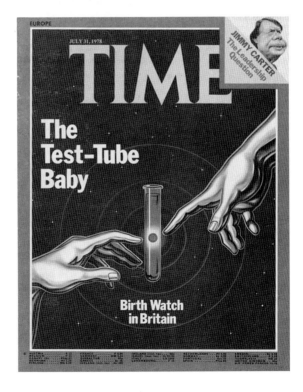

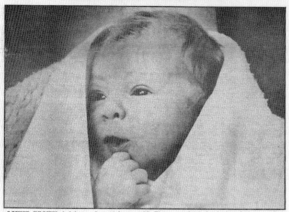

Daily Mail

THURSDAY, JULY 27, 1978   8p

WORLD EXCLUSIVE

And here she is...
THE LOVELY LOUISE

LOUISE BROWN, bright-eyed at 18 hours old: The test tube baby in hospital yesterday

Daily Mail World Exclusive Picture by Bill Cross ⓒ World Copyright Associated Newspapers Group Ltd., 1978.   Full story and more pictures inside

15. Thursday, 17 July 1978. A perfect baby . . . and a perfect front page. The *Daily Mail* making the most of its exclusive access to Louise Brown, and underlining that the product of this technology was a new person.

16. The *Daily Express* offering a different treatment of the same event. Note the identity of the people produced with the Bernard's bedroom chemistry set. He has plainly been cloning himself.

'Daddy has warned you, Bernard—if you don't stop, he'll take your chemistry set away.'

17. The *Boston Globe*'s view of the response at MIT to the decision by the lay members of the Cambridge Experimentation Review Board to allow recombinant DNA experiments to continue in the city.

"Crack out the liquid nitrogen, dumplings ... we're on our way."

18. Frankenstein continues his effort to persuade biologists to accept parental responsibility, this time in a British cartoon from the *Guardian*, which accompanied an article on regulations for genetic manipulation.

19. The intermingling of species has changed little between (a) sixteenth-century monsters and (b) the creatures in this 1980s German cartoon laboratory.

20. Frankenstein as the modern Gulliver of gene technology, tied down
by the regulators in Germany in 1987.

21. The interior of the body has altered out of all recognition since Vesalius. The self-anatomising figure drawn for the *Guardian* in 1996 displays a bar code, the DNA readout that is the key to his outward form and his innermost character.

However, the rise of molecular biology was only part of the story of the popularisation of new powers derived from the life sciences. While the new field might be a distinct enterprise to scientists, for outsiders it was just one among several strands, some of them much more traditional, that were woven together to create a new tapestry. The 'secrets of life' which were being uncovered were not just about the workings of DNA. Molecular biology was a scientific revolution, by any standard, but the new popular notion of a 'biological revolution' had more complex connotations.

The term entered wide popular usage toward the end of the 1960s, a few years after genetic engineering. Part of its appeal undoubtedly lay in its ambiguity. At least three uses co-existed. First, there was the revolution in understanding, a fundamentally new departure in scientific knowledge (although not a scientific revolution in Kuhn's more specialised sense).[44] Secondly, there was the idea of a technological revolution, with a transformation of economic infrastructure and social and industrial organisation – akin to the industrial revolution or, some suggested, the computer revolution. Finally, there was the analogy with political revolution – conceived and engineered by a group of individuals working to realise a shared vision of the future.

Scientists and popularisers of science were referring to the revolution *in* biology from the start of the 1960s. The first chapter of Kendrew's *Thread of Life*, for example, bears that title. The second meaning was more often employed by popular writers of the later 1960s. They combined the older conception that the successor to the industrial revolution would be a biological revolution, an idea going back to the 1920s and 1930s,[45] with the sense that revolutionary changes were coming over biology itself. Molecular biology would thus become the means to realise these much older expectations. It would be Huxley's 'really revolutionary revolution'.

The third, quasi-political sense, was less common, but was occasionally used to denote the programme of those dubbed 'biological revolutionaries'. As the decade progressed, these different, sometimes overlapping meanings blended to generate an idea of a more expansive biological revolution. For some, the blending was self-conscious, as for the historian Donald Fleming in his lengthy article in *Atlantic Monthly* in 1969, 'On living in a biological revolution'. Fleming began by listing twelve new discoveries, including those of molecular biology but also work on contraception, *in vitro* fertilisation, immunology, pharmacology and transplant surgery. His contention was that 'Each of these is a major discovery or complex of discoveries in itself, but *they add up to far more than the sum of their parts*. They constitute a veritable

Biological Revolution likely to be as decisive for the history of the next 150 years as the Industrial Revolution has been for the period since 1750.'[46]

In the following paragraph, though, the ground of the metaphor shifts, and Fleming suggests that 'every full-scale revolution has three main components: a distinctive attitude toward the world; a program for utterly transforming it; and an unshakeable confidence that this program can be enacted – a world view, a program, and a faith.' In this light, Fleming's first 'biological revolutionary' is Müller, the revolutionary socialist and eugenist. Of the rest, all but one are geneticists or molecular biologists. 'The most conspicuous representatives of this revolutionary temper in biology', according to Fleming, are Joshua Lederberg, Edward Tatum and Francis Crick. The exception is the British physiologist Robert Edwards, whose work I discuss in the next chapter. He is 'clearly in training for the revolutionary cadre'.

The goal of these social revolutionaries is 'the manufacture of man'. They believe that our only salvation lies in a programme including 'control of numbers by foolproof contraception; gene manipulation and substitution; surgical and biological intervention in the embryonic and neonatal phases; organ transplants or replacements at will'. Their revolutionary fervour is depicted with dry distaste. Lederberg's proposals for euphenics – modification of development – and his interest in embryology draw the comment that 'he is at once maddened and obsessed by the nine-month phase in which the human organism has been exempted from experimental and therapeutic intervention – such a waste of time before the scientists can get at us'.

Just as important as this style in establishing their revolutionary credentials, however, is the genuine revolution in biological understanding. The switch to this sense of the word is accomplished by an adroitly ambiguous use of 'culture', to denote both a set of values and rules of conduct and a research culture. 'One of the stigmas of revolutionaries in any field is their resolute determination to break with traditional culture,' Fleming writes.

> For a scientist, the most relevant definition of culture is his own field of research . . . Today, the pretence that biologists somehow constitute a community has been frayed to breaking point. At Harvard, for example, the revolutionaries have virtually seceded from the old Biology Department and formed a new department of their own, Biochemistry and Molecular Biology. The younger molecular biologists hardly bother to conceal their contempt for the naturalists, whom they see as old fogies obsequiously attentive to the world as it is rather than bent on turning it upside down.

This deliberate depiction of the scientific revolution as analogous to a political revolution also dramatises the way the achievements of modern biology may threaten traditional values and institutions outside science. But Fleming's article shows, too, how other kinds of science were needed to sustain his construction of a wider biological revolution. Reliable contraception and transplant surgery were important exhibits. They were practical, first of all. And transplant surgery, in particular, was the kind of feat that had been used to dramatise the potential power of biology for almost a century. The idea of a biological revolution in the technological sense demanded that biology be applied to human beings. But molecular biology, as yet, was a revolution in knowledge only, and a knowledge expressed in very abstract terms which bore little obvious relation to everyday life. The manipulative possibilities it might offer were hard to grasp until they might actually be put into practice – something which, so far as human genetic alteration goes, we are still waiting for.[47]

So the revolution in knowledge became just part of a picture which was filled out with concrete, practical achievements derived from other areas of the life sciences. And there were many of these in the 1960s. For wider publics, the biological revolution was as much about the pill, heart transplants, kidney machines, foetal surgery, hallucinogenic and antidepressant drugs and *in vitro* fertilisation as about the 'secrets of life' encoded in DNA. These new applications broadened the terms of the notion of a biological revolution and dramatised its potential impact.

Some of the most striking demonstrations of this were in magazine serials, a more important medium then than now. Here, the use of lavish colour illustration rivalled the visual impact of television, while printed text was less ephemeral than a TV commentary. A revealing example is the four-part feature in *Life* in 1965 under the general heading 'Control of life'. The research, chosen for its photographic possibilities, produced an extremely vivid series of images. Consider the first article in the series, on embryology. We see a monkey foetus lying in a surgeon's hand, and opossum embryos in an incubator. The largest image is a double-page spread showing E.S. Hafez, then working on cryopreservation, acting out his image of the future. There are five vials in close-up, balanced on Hafez's outstretched palm, at the end of his white-coated arm. Each vial carries a label – swine, sheep, cattle, rabbit, man. The caption dips into science fiction: 'Cold storage embryos to send to distant planets as a test-tube colony'.[48] After this graphic beginning, the series continued with a kidney transplant operation and a piece on human spare parts.[49] Only in the final article, by *Life*'s science editor Albert Rosenfeld, were the prospects for genetic engineering discussed. This piece recalls others already mentioned, but its impact was undoubtedly strengthened by the technologies already displayed.[50]

The most widely discussed of these technologies were those used in organ transplantation, particularly heart transplants after Barnard's first operation on Louis Washkansy in December 1967. The ancient view of the heart as the seat of the spirit made this a potent symbol, and the flurry of 105 heart transplants in 1968, dubbed the year of the transplant by the press, fuelled enormous publicity in many countries, each with its own series of patients. Even in 1969, Fox and Swazey report that subscription to a cuttings service between January and May produced no fewer than 2,215 clippings on organ transplantation and artificial organs from American newspapers. They found that 'the transplantation of a human heart was depicted as the quintessence of the "daring adventure" that medical research could be, and it was equated with the feats of the astronauts who made the first manned lunar orbit in 1968'.[51]

They also comment on the feelings expressed by some patients about kidney machines, another medical innovation which entered widespread use in the mid-1960s.

> For some patients, 'seeing my blood outside of my body running through coils of synthetic tubing' is deeply distressing. Many regard the machine in an anthropomorphic way, as a 'miraculous . . . powerful monster . . . with an almost frightening hold on my life', a 'rigged roulette wheel', reducing me to a 'half-robot, half-man'.[52]

Some of this ambivalence toward the machine also percolated into press reports of dialysis programmes, particularly when selection criteria for patients were discussed. The general interest in medical developments like this was also reinforced by events such as the first (and at the time only) implantation of an artificial heart in April 1969, and an attempted human eye transplant in the same month, both in the US. Controversy attended both operations, and both the surgeons concerned were eventually censured by their colleagues.[53]

Through these and other developments, the notion of the biological revolution, beginning with molecular biology but also gathering in other, more easily understandable feats, became established. The range of ideas implied by the term, and the kind of unease they evoked, are well represented in a widely read popular book of the time, Gordon Rattray Taylor's *The Biological Time Bomb*, published in 1968. Examining Taylor's book, and some of the responses it evoked, gives a measure of the distance in public debate spanned by the decade between Beck and Rostand's books and this new look at the social implications of biology.

Taylor's book was a review of the full range of biological discoveries of the 1960s which had caught the public eye, and of the expressions of concern that had often greeted them. Like Fleming, he argued that they were more than the sum of their parts, and he tried to advance the discussion as well, to get beneath the abstractions and spell out in detail some of the consequences of biological developments in several areas. It was a pessimistic book, taking some of its rhetoric from the first wave of environmentalism, if not yet from the 1960s counter-culture.[54] It was seen at the time as something of a landmark in public debate. As one sympathetic British biologist put it, it was 'the first major exposition, addressed to the general public, of issues which are going to be very much with us during the next few decades'.[55] In his use of arguments and anecdotes from earlier works, his selection of experiments for discussion and choice of past biological innovations as clues for thinking about the future, Taylor produced a text which, rather like Haldane's *Daedalus*, both exemplified the reasons why biology came to be a matter of public concern, and indicated some of the directions in which that concern would develop.

The Biological Time Bomb was distinctly different from the numerous popular expositions of the new biology already published, although many books similar to Taylor's would follow.[56] He had little concern with the details of the science as systematised knowledge, and there is none of the hierophantic tone of some works by scientists. It is the work of a competent professional journalist, selecting developments of immediate relevance to his theme, discarding the rest. Taylor wove together a large number of accounts of research work, with just enough scientific background to make them intelligible, to make a picture of a biological revolution in the broadest sense.

We are now, though we only dimly begin to realize the fact, in the opening stages of the Biological Revolution – a twentieth century revolution which will affect human life far more profoundly than the great Mechanical Revolution of the nineteenth century or the Technological Revolution through which we are now passing.[57]

The advances which substantiate this are those already considered here. After his first chapter ('Where are biologists taking us?'), there are detailed discussions of likely progress in reproductive physiology (Chapter 2: 'Is sex necessary?'), transplant surgery (Chapter 3: 'The modified man'), age research (Chapter 4: 'Is death necessary?'), and neurochemistry (Chapter 5: 'New brains for old'), before genetic engineering is discussed in Chapter 6. Successes already recorded are shown to have created social, legal and ethical problems; Taylor outlines these to 'give us a better idea of the amount of confusion which could be caused by some of the other possibilities described'.[58] There is an extended account of legal and ethical issues arising in cases of organ transplantation, for example. And in a discussion of the pill, Taylor writes, 'The discovery of oral contraceptives is only the *beginning* of a new world of expertise in the control of the reproductive process, and the controversies which have raged over the pill will soon be lost in the thunder of even more desperate battles.'[59]

On the other hand, the growth of molecular biology is used to underwrite the revolutionary character of the changes coming over science. And it is molecular biologists who are most commonly quoted on 'the direction in which biology is heading'. The opening chapter alone assembles quotes from Joshua Lederberg, Bentley Glass, Francis Crick, Salvador Luria and W.H. Thorpe, all of them emphasising the sweeping changes now in prospect and the fears they may raise. All but the last-named were molecular biologists. Similarly, in Chapter 6 Taylor claims that 'To judge from what the scientists themselves are saying, the most serious of all the human problems created by biological research is constituted by man's imminent power to interfere in the processes of heredity to alter the genetic structure of his own species.'[60]

Most of the quotes he uses come from essays and addresses published in the 1960s, as do the bulk of the experimental examples. But Taylor also makes use of earlier sources to set these in context. Although a number of specialist symposia had discussed issues raised by biological advances[61] – and he drew on these freely – his was the first book on the topic addressed to a popular audience since Rostand, and much of the material the Frenchman used reappears in *The Biological Time Bomb*. Indeed, the substance of both the first paragraph of the book – describing an episode from Diderot's *The Dream of D'Alembert* which appears to prefigure ectogenesis – and the very last – with its quote from Francis Bacon – is also to be found in Rostand's book.[62] Time and again, anecdotes and quotations from Rostand, sometimes acknowledged, sometimes not, are used to point up a particular issue, summarise trends, underline Taylor's own observations or give an account of recent events an historical dimension.[63] In some ways, Taylor simply took up the framework of the earlier book and extended it to incorporate new discoveries in the ten years since. But the many similarities should not obscure the equally significant differences between the two.

One is the much greater number of examples of successful biological control Taylor had to draw on. His book is much longer than Rostand's, but free of repetition. Taylor's book is a partial record of an extraordinarily productive decade of experimentation, far surpassing anything which had gone before in concrete success in gaining control over biological systems. There are few traces in the later book of Rostand's frequent tentativeness in prediction – Taylor has absorbed the confidence of the biologists.[64] His provision of such a wealth of examples goes some way towards answering Barnett's criticism of *Can Man Be Modified?* in his 1959 review. He also makes good a second major omission, with his more detailed consideration of real consequences of biological innovation, and a determined effort to spell out specific possibilities. Perhaps the most striking difference between the two, however, lies in their attitude to restraints on scientific knowledge. Rostand's question, whether 'science [is] reaching a frontier beyond which its progress might be more harmful than advantageous' is now answered in the affirmative. For Taylor, 'apart from the actual misuse of new powers, whether by accident or design', there is a further question: 'How great a rate of change can society stand?' His answer is unequivocal:

> It seems quite possible that the rate of biological innovation may be so high as to destroy Western civilization, perhaps even world culture, from within, creating a disoriented, unhappy and unproductive society, unless it is brought under deliberate control.[65]

Here speaks the authentic voice of the fear of the final phase of modernity, perceiving that all that is solid may melt into air, and seeking to call a halt. But the weakness of the argument becomes apparent in Taylor's vagueness about how to bring research under control. He speaks about the need to 'devise adequate institutions to take broad social decisions' but his suggestion lacks conviction. Indeed, by his last chapter, he is arguing that the profit motive will drive research forward and, more generally, that 'current indications are that the world is bent on going to hell in a handcart, and that is precisely what it will do'.[66] It is a curiously unsatisfactory verdict, more suited to a writer of fiction. If this is what the author believes, why bother to write at such length about recent and possible future science? Why argue that scientists have a duty to inform the public about their work and the ethical and social issues which may arise from its application? Perhaps there is still hope. But whether it lies in social adaptation to biological innovation, in control of research, or whether neither of these is possible, is left unresolved.

An unsettling book, then, which attracted wide attention, aided by a BBC television programme on *The Assault on Life* shortly before it was published. Most reviewers of the book agreed with what they took to be Taylor's position – that researchers could not be hindered but some effort must be made to control application of new findings. One dissenting reviewer, in the *Times Literary Supplement*, assumed that the book was a call for restrictions on research itself. The anonymous writer dismissed almost all of Taylor's speculation as mere science fiction:

> If Mr. Taylor's real motive is . . . to promote serious and informed public discussion . . . he does his case no good by confusing fact with fancy and by treating both as if they were equally plausible . . . Sensible public discussion is certainly not furthered by books with sensational titles prophesying doom.[67]

In any case, in the *TLS* writer's view, biology is already under social control, for 'in all highly developed countries elaborate machinery exists to determine and control expenditure on scientific research; and even the smallest projects are carefully scrutinised by competent and socially responsible bodies'. He concludes that 'society cannot, and should not try to, decide what a man may discover, but society can and does decide how any discovery is to be applied'.

This dismissal thus argued that the bulk of Taylor's book was a work of fiction, a rhetorical ploy I will examine in more detail in relation to 'genuine' fiction in the next chapter, and that the established institutions of science policy were adequate to control the research. But this was untypical. Most responses, on both sides of the Atlantic, were well in keeping with Taylor's more apocalyptic mood, and his feeling that modern societies were generating the seeds of their own destruction by driving science into new areas of life. Psychiatrist Anthony Storr set the tone in the *Sunday Times*, writing that 'the discoveries of modern biology may be as threatening to man's survival as the hydrogen bomb'. Biologist Robert Morison wrote in the *New York Times*, 'the overall picture of an explosive increase in biological knowledge that will require what amounts to a revolution not only in individual morals but in political, economic and social thinking is undoubtedly correct'. Critic George Steiner also endorsed the idea of a biological revolution, commenting that 'responsible observers' saw in this revolution 'the first total crisis in the history of the species'.[68] And there were similar views to be found in the *Economist*, *New Scientist, Science Journal*, the *New York Review of Books* and the *New Statesman*.[69]

The reviewers also agreed that science should not or could not be regulated, thus amplifying the ambiguity of the book, but they did not see this as Taylor's intent. Morison felt that 'the purpose of the book is clearly to stimulate the acquisition of more knowledge, rather than less', a view shared by Nigel Calder. Storr affirmed that 'the progress of science cannot be halted, even if it were desirable to do so'. This shared opinion, a rather weak response to a 'total crisis', caused several reviewers to echo Taylor's pessimism. Steiner concluded that it was indeed quite likely that the developments Taylor outlined would lead to disaster, but that, if so, we were already too far along the technological road to take any effective action to prevent the catastrophe. Others spoke of the book's 'morbid fascination', its 'blend of compulsive interest and horror' as though it were indeed a Gothic novel.[70]

The collected reviews tell us something about the temper of the late 1960s – in between Rachel Carson's *Silent Spring* of 1962 and the Club of Rome's *Limits to Growth* in 1972. But while doomsaying was undoubtedly widespread among intellectuals in those years, they generally resisted extending the idea of limits to growth to include the possibility of limits to knowledge. Most reviewers treated the book a good deal more seriously than was the case for Rostand's volume ten years earlier. Nevertheless, these commentators, mostly senior scientists or long-standing science writers, shared Taylor's commitment to freedom of inquiry. They tell us little of the general public reaction to the book, or to the growth in the powers of biology. To see that in more detail, after this overview of the first two post-war decades, we must look more closely at public debate about some specific areas of research.

Chapter Eight

The Baby of the Century

Man is the watershed that divides the world of the familiar into those things which belong to nature and those which are made by men. *To lay one's hands on human generation is to take a major step toward making man himself simply another of the man-made things.* Thus, human nature becomes simply the last part of nature which is to succumb to the modern technological project, a project which has already turned the rest of nature into a raw material at human disposal.

Leon Kass, 1972, pp. 49–50; original emphasis

If ever a research programme drew on fictional images for its inspiration, and interpretation, it was the work which led to human *in vitro* fertilisation. The *Frankenstein* script has been generalised to apply to almost any technology, even though it still has a special affinity with technologies of life. The idea of conception outside the body identified with *Brave New World* has a more specific connotation. And once *that* script established its hold, any research which seemed to offer control over reproduction was readily figured as a step towards Huxley's world.

A glance at reports of work on human *in vitro* fertilisation confirms that it was almost always described as part of an effort to produce a 'test-tube baby'. The step-wise nature of the research made the saga of *in vitro* fertilisation one of the main long-running news narratives about experimental biology for ten years or more. Primed by writers such as Huxley, Rostand and Taylor, journalists seized on the story as a symbol of the wider biological revolution heralded by the advent of molecular biology.

160

The manipulation of eggs and sperm in the test-tube (actually, the petri dish) was nicely poised between old and new. It was in some ways biology at the level at which Jacques Loeb worked in the early years of the century and, like his achievement of artificial fertilisation in frogs, was heralded as the creation of life in the laboratory. This version of creating life now co-existed with the newer notion that the secret of life lay in molecular biology, and life creation meant DNA synthesis. At the same time, it was reasonable to assume that successful *in vitro* fertilisation was a prerequisite for any attempt at genetic engineering by manipulating the DNA of the germ cells. Similarly, successful embryo implantation would be a necessary stage in any notional procedure for human cloning, at any rate pending full extra-corporeal gestation.

Aside from these associations, there was the fact that the work was advanced enough to produce results, and images of those results. At first, these were photomicrographs of a sperm fusing with an egg, but eventually there were real babies to photograph. Between the end of the 1960s and the end of the 1970s the procedure for implantation and growth to term of a human egg fertilised in the laboratory shifted in status, from a future possibility among many to one that had actually been realised in practice. Before moving to that decade, though, some account is needed of the origins of the modern work, to relate developments in this field to more general trends in biological research.

Reproductive physiology becomes an applied science

Producing a test-tube baby has three technologically assisted stages. The mother must supply an egg or eggs at the correct stage in development; one or more eggs must be fertilised; and the fertilised egg must be implanted in the womb. After that, normal gestation proceeds, though still under a technologically intense regime. The whole procedure requires strict management of a host of hormonal, biochemical and other environmental conditions. The gradual definition of these conditions, and development of techniques for controlling them, resulted from efforts in many different areas of physiology and biochemistry, efforts often directed to other ends. Overall, work on *in vitro* fertilisation was sustained mainly by the emergence of reproductive physiology as an applied science in two areas: human contraception and stockbreeding.

In the 1960s, all of these developments were relatively recent. Although the idea of ectogenesis was popularised in the 1920s, little scientific work had been done at that time that was relevant to even the first stages of fertilisation. The reproductive hormones were unknown before 1920, and most of our knowledge of the fine structure of sperm and egg dates from after World War II. Much of our knowledge of how the gametes behave was built up by those who drove development of the oral contraceptive pill, notably Gregory Pincus, his colleague M.C. Chang and John Rock.[1]

Pincus, who as I noted earlier had a troubled academic career because of the public notoriety of his work on fertilisation in rabbits in the late 1930s, nevertheless became an authority on reproductive endocrinology and the mechanism of fertilisation. He was co-founder of the Worcester Foundation for Experimental Biology, in Massachusetts, in 1944, and recruited Chang from Cambridge the following year.[2] Pincus was also a consultant to G.D. Searle and Co., who were shortly to duplicate Djerassi's work on the synthesis of an oral progestogen. Chang and Pincus began seriously investigating oral contraceptives in the early 1950s following a meeting between Pincus, the birth control campaigner and eugenist Margaret Sanger and a wealthy supporter of Sanger's, who agreed to pay for the research. Post-war alarm about a 'population explosion' – one of the early harbingers of the 'limits to growth' debate – helped speed up a shift in attitudes toward scientific research into contraception as the pill became a practical possibility. A report in *Science Newsletter* of a 1951 conference on population problems attended by Pincus was headed 'The explosive increase in world population could be squelched by a tiny pill'.[3]

Pincus and Chang later collaborated with Dr John Rock, professor of gynaecology at Harvard, who had been working on the uses of hormones in female infertility. In the late 1950s the three supervised human trials of a synthetic progesterone pill, first on female psychiatric patients in the USA and later on a larger scale among women in Puerto Rico. The G.D. Searle formulation Enovid was approved for sale in the USA in 1960. Two years later almost two million American women were regular users.

How could the control over reproduction displayed in the pill be turned to promoting conception? *In vitro* fertilisation would be one way. Although Pincus had published claims to have achieved artificial parthenogenesis in an isolated rabbit egg as early as 1934, attempts to repeat the experiment failed. Similarly, John Rock claimed to have recorded *in vitro* fertilisation of human eggs in 1940. These and later claims by fellow American Landrum Shettles – whose publication of photomicrographs which he read as showing *in vitro* fertilisation provoked condemnation from the Catholic Church – were all strongly criticised on scientific grounds by Chang. He maintained that these results were more likely due to spontaneous division of degenerating eggs.

Throughout the 1950s, Chang worked to refine understanding of sperm and egg interactions. He made it clear that fertilisation must be considered as a process, not an event, and defined capacitation – the stage a sperm must pass through before fertilisation can take place. He also investigated the developmental paths followed by the egg from release in the ovary to possible implantation in the womb after fertilisation. In 1959, Chang reported the first mammalian birth after *in vitro* fertilisation. He fertilised a rabbit egg in the laboratory using capacitated sperm recovered from the uterus of a female rabbit killed just after mating. The resulting blastocyst was implanted in another female and brought to term. The essential stages for human *in vitro* fertilisation to be feasible had now been defined.

Meanwhile, the technology was also progressing rapidly in stockbreeding. Artificial insemination had long been used on a small scale in cattle, and following Parkes's discovery in 1949 that addition of glycerol enabled sperm to be frozen without damage, it expanded very rapidly. But this accounted for only half the genetic material of the offspring, and researchers then turned their attention to increasing the number of eggs produced by prize cows. One possibility seemed to be embryo implantation and by 1957 a British group at the Agricultural Research Council laboratories near Cambridge had transplanted naturally fertilised sheep's eggs from one ewe to another. Again, this demanded detailed knowledge of mammalian reproductive physiology, especially to synchronise the cycles of donor and recipient.

At this point, developments in human and animal research diverged somewhat. *In vitro* fertilisation was of little interest in stockbreeding, and here efforts concentrated on producing large numbers of eggs by hormone-induced superovulation, which would then be fertilised in the mother. The multiple embryos produced would then be flushed out of the donor's womb and implanted in other females for full gestation.

Thus, by the end of the 1950s, successful applications of past work in reproductive physiology were on the verge of large-scale use, and the scientific discoveries that had made these applications possible furnished a secure foundation for concentrated work on human *in vitro* fertilisation in the following decade. The early reception of that work was in the context of the emerging idea of a biological revolution.

How sinister is biology?

Huxley's novel and Rostand's speculations gave the image of the 'test-tube baby' a prominent place in the prophecies of a biological revolution. In 1968 Taylor cited a large number of experiments in the field in support of his contention that 'biological science is taking the reproduction process apart in a far more thorough way than is yet generally understood'. This was his first major piece of evidence for a biological revolution.[4] At the time it was based on a range of reports throughout the decade. Like other projects which became symbols of the manipulative potential of biology, the early history of *in vitro* fertilisation research abounds in premature, exaggerated or unsubstantiated announcements, which were taken up widely by the press only for their authors to fade into obscurity.

In January 1961, for example, less than two years after Chang reported his *in vitro* rabbit, Dr Daniele Petrucci, a University of Bologna physiologist, let it be known that he had fertilised a human egg *in vitro* and grown the resulting embryo for twenty-nine days in the laboratory, at which point he destroyed it because it became deformed.[5] The work evoked a violent reaction from the Catholic Church, which apparently led him to abandon the research. But although no details of Petrucci's methods or results appeared in any scientific journal, his claims entered the popular histories of experimental biology. Similar descriptions of his work appear in several much later books and articles.[6] All refer to Petrucci's subsequent, somewhat mysterious visits to the Moscow Institute of Experimental Biology, suggesting that the Russians might have been pursuing such studies in secret.

In the West, Haldane and Birkenhead's predictions of adverse reaction were amply fulfilled. Catholic critics were fierce. *America*, a leading US Catholic journal, suggested that 'The spirit of Frankenstein did not die with the Third Reich. His blood brothers regard a human being as just another expendable microbe, provided it is legally defenceless, physically helpless, and tiny enough to ride on the stage of a microscope.'[7] A report from China, though, suggested a polarisation of views. There, the Communist Party newspaper commented that 'These are achievements of extreme importance, which have opened up bright perspectives for similar research . . . If children can be had without being borne, working mothers need not be affected by childbirth. This is happy news for women.'[8]

So those who were striving to make a new society were prepared to remake reproduction in the process, while more traditional voices evoked the twin spectres of Frankenstein and Nazism to warn against doing so.

Petrucci's experiment appeared to suggest that some were working with human material. There were few if any reports of work on human gametes during the next six or seven years, but the climate of expectation grew stronger. Wider use of human artificial insemination by donor to overcome some cases of male infertility seemed to provide a precedent for manipulation of germ cells. Schemes for eugenic improvement still received occasional publicity, though they were generally poorly received. Hermann Müller, for example, repeated his proposals for 'germinal choice' in the late 1950s, after keeping silent on the subject for some years after the war. Now, his proposals for artificial insemination with the preserved sperm of individuals with desirable traits drew frequent criticism for their authoritarianism, poor definition of the qualities to be promoted, and inconsistency with some of his earlier suggestions.[9] It was pointed out that he now omitted Lenin from his list of worthy donors, for example.[10] In 1962, Julian Huxley extended Müller's scheme to include use of implantation of fertilised ova, in his Galton lecture of that year on 'Eugenics in evolutionary perspective'. When reprinted in his volume *Essays of a Humanist* two years later, the proposal was criticised by one reviewer for 'arrogance, historical naiveté and . . . authoritarianism'.[11]

Meanwhile, studies in animal physiology were proceeding apace. In the early 1960s they produced a number of scientific stunts. That most often quoted was staged by Rowson in 1962, and involved the air freight of fertilised ova of one breed of sheep from Britain to South Africa, carried in the fallopian tubes of a live rabbit. On arrival, the eggs were removed and surgically implanted in the wombs of ewes of a different breed. The resulting Border Leicester lambs borne by Dorper ewes were an impressive demonstration of scientists' growing control over the reproductive process, and their power to dissolve old boundaries of time, space or species. As Taylor put it, it was 'something recalling the kind of wild rumours which used to circulate in the dark ages'.[12]

Another distinguished animal biologist, E.S.E. Hafez, featured in the *Life* photo-spread described in the last chapter. In the same article, he predicted that embryo freezing techniques would soon be applied to people. Several subsequent authors quoted his view that

> A woman will be able to buy a tiny frozen embryo, take it to her doctor, have it implanted in her uterus, carry it for nine months, and then give birth to it as though it has been conceived in her own body. The embryo would, in effect, be sold with a guarantee that the resulting baby would be free from genetic defect. The purchaser would also be told in advance of the colour of the baby's eyes and hair, its sex, its probable size at maturity, and its probable IQ.[13]

This is the scenario now repeated many times under the heading of the 'genetic supermarket', or more recently as procuring 'designer babies', already on offer in 1965. As ever, only a few scientists have to support this kind of speculation for it to become part of a wider picture of what their research is all about. And this time, a fresh ingredient was soon added – concrete progress towards human applications. The climate of expectation was reinforced when the first fully documented success in human *in vitro* fertilisation – without implantation – came in February 1969. The two British researchers involved, Patrick Steptoe and Robert Edwards, received enormous publicity then, and again a year later.

The work centred on the physiology department at Cambridge University, where Edwards began working with the professor of physiology, C.R. Austin. Edwards moved to Cambridge because the Medical Research Council refused to support his plans to switch his studies of mammalian fertilisation from pigs to humans. In Cambridge there was support from the Ford Foundation for basic research into fertility, from a fund set up to support development of new contraceptives.

In the mid-1960s, Edwards began a long collaboration with Patrick Steptoe, a gynaecologist working in Oldham. Steptoe's introduction of laparoscopy promised a steady supply of ripe ova, which could now be harvested via a small incision in the abdomen. Even before this new technique gave fresh impetus to his work, though, it was already clear that Edwards had set his sights on human *in vitro* fertilisation. An early article describing his efforts in the *Guardian* in 1966 referred to experiments performed on human oocytes obtained during surgery and suggested that 'it is now reasonable to suppose that in the next decade it could be possible to fertilise oocytes in the laboratory and artificially nurture them as developing embryos'.[14] This is what led

Fleming to dub Edwards one of his 'biological revolutionaries' in *Atlantic*. In the same month as Fleming's article appeared, Edwards and Steptoe announced their first successful human *in vitro* fertilisation.

The issue of *Nature* for 15 February 1969 included a paper entitled 'Early stages of fertilisation in vitro of human oocytes matured in vitro'.[15] This was before the days when that journal produced a weekly press release, but the paper promised more in its abstract, saying that 'Human oocytes have been matured and fertilised by spermatozoa *in vitro*. There may be certain clinical and scientific uses for human eggs fertilised by this procedure.' And *Nature* drew further attention to the paper in a lengthy editorial.[16] The writer tried to dissociate the work from the images it would evoke. In one of those texts which reinforces the very associations it tries to deny, it was suggested that 'The day of the test tube baby is not here yet, and the advantages of this work are clear. These are not perverted men in white coats doing nasty experiments on human beings, but reasonable scientists carrying out perfectly justifiable research.'

The first responses elsewhere in the press suggested that *Nature's* argument might carry the day. *The Times* took its account from the *Nature–Times* news service, which offered a simplified factual account of what had been achieved in Cambridge, closely based on Edwards's paper. The piece was, however, promoted to the front page and given the headline 'New step towards test tube babies'.[17] A similar headline appeared in the *Daily Mirror*, which followed *Nature's* lead in emphasising the benign intentions of the scientists.[18]

But despite these reassurances, the idea that babies as opposed to blasto-cysts were about to be grown in the laboratory was never far away. The *Guardian*, for example, followed a straightforward account of the experiment, under the heading 'New hope for the childless', with a quote from the previous evening's Independent Television *News at Ten*:

> There could be an even further stage of development – with untold implications for the whole science of medicine – the scientists are now in a position not only to replace the fertilised egg in the mother's body, but to continue to develop the fertilised egg artificially and perhaps even to produce a human baby without using the mother's body again at all.[19]

The next day, the paper's science correspondent gave a fuller account of Edwards's work, very similar to the one he gave three years earlier, but now with an opening comment that it was quite misleading to describe the experiments on fertilisation as 'the creation of life in the test tube'. The logic which applied harked back to the debates about Loeb's amphibians: 'Life, or at least the potentiality of life, resides in the oocyte and the spermatozoa; the artificial creation of either of these is so far beyond the present horizon of knowledge that it can be discounted as a serious possibility.'[20]

The *Guardian's* cartoonist, though, was not persuaded. The article ran alongside a series of frames which went to the heart of the fears of Faustian or Promethean experiments which the paper's science writer was trying to soothe away. In the first frame, a white-coated scientist appears, having just produced a tiny baby – a recognisable homunculus rather than a realistic foetus – which sits inside a small test-tube. It is the image evoked by the journalistic phrase of the moment. In succeeding frames, the baby grows, emerges from his glass prison, and develops into a huge and overpowering grotesque. In the final frame the roles are reversed. The scientist is now helplessly corked inside the tube, looking up at his creation, pleading: 'Son! Let me out son, listen to me son! SON!'

Perhaps this series of images, in conjunction with the text, was intended as an ironic comment on foolish fears of the kind *Nature* alluded to.[21] But such efforts are equally open to more literal readings. This effect also appears in other newspaper discussions. *The Times*, for example, underlined the significance of the reported fertilisation by running four follow-up articles the day after the first front-page report. First came a report of a London press conference given by Steptoe and Edwards, stressing that 'the hope was to cure infertility. There was no question of mass-producing test-tube babies.'[22] This piece was supplemented by an article credited to the *Nature–Times* news service expanding on the scientific details, headed 'Implications of the the test-tube *embryo*' (my emphasis), although quoting a fellow scientist to the effect that the test-tube baby was 'pure science fiction'.[23] On the same page a separate article reported a sample of non-scientific opinion on Edwards's work, giving prominence to opposition from the Catholic Church. Most of the other organisations and individuals invited to comment were positive, however, especially about the prospects for alleviating infertility.[24]

A very different view appeared in the fourth item, a leader article headed 'Life in the test tube'. This began by suggesting that 'It is the implications rather than the direct consequence of the first successful fertilisation of a human egg in a test-tube that pose the most searching moral problems.'[25] One of the implications was that procreation from a single cell might one day be possible, and 'once it is, it is this which will provide the real challenge to all our assumptions about the reproduction of the human race'. In particular, it would permit effective selective breeding. 'It is not fanciful to say that at this stage, which has of course not yet been reached, the end of human beings as a wild breeding race could be in sight.' This eugenic interpretation of *in vitro* fertilisation was a 'frightening thought', because the consequences were so unpredictable. For example,

> the cheapest and surest way for any small, impoverished country to improve its wealth and influence would be to concentrate on breeding a race of intellectual giants . . . and as soon as one nation adopted a policy of effective selective breeding . . . others might well feel compelled to follow suit. The threat that this might pose to accepted human values would be extremely grave.

The editorialist concludes that 'there can be no slick answers to this challenge, but it is not too soon to begin considering the nature of the problem'. As with earlier, more general discussions of genetic engineering, restraints on research seem not to be on the agenda, and the reader is left with the novel problem of a eugenic arms race to worry about, without any obvious solution on offer.

The idea *The Times* chose to declare 'not fanciful' had been aired the previous year in a well-regarded science fiction novel by Briton John Brunner, *Stand on Zanzibar*. The main plot element in this long, complex and ambitious book was the attempt of the Far Eastern republic of Kakatang to develop a superior population through ectogenesis, and the dangers that this posed for an uneasily balanced, overpopulated near-future world. Brunner created his world-picture from a collage of materials, real and imagined, intercalated with episodes of the narrative in the manner of Dos Passos. The background to the action was thus very detailed, but Brunner indicated that the idea of a breakthrough in ectogenesis was central to his construction of the book, 'because no other event could cause such a dramatic upheaval in the kind of world I was thinking about'.[26]

Brunner made clear that his novel was, in part, a comment on contemporary scientific developments by including in the text an item on cloning from the front page of the London *Observer* of November 1966, 'New Einsteins from Cuttings'. The news report, in turn, was based on an article by Joshua Lederberg in the *Bulletin of the Atomic Scientists*.[27] *The Times*, in effect, then endorsed Brunner's fictionalised view of the implications of ectogenesis. Both articles helped make *in vitro* fertilisation a touchstone of the biological revolution.

This time, the coverage tailed off fairly rapidly, merely priming audiences for the greater success which would come nearly ten years later. Even so, there were some personal expressions of disquiet at the direction biology was taking. A correspondent in *The Times*, for example, compared the first *in vitro* fertilisation with the first nuclear explosion, and spoke of a 'piercing feeling of regret'. He continued:

> Scientists, particularly medical scientists, must often be caught in the Promethean dilemma between seeking new knowledge on the one hand and, on the other, of respecting that which is forbidden to them. In this case is it unfair or obscurantist to suggest that they have gone too far?[28]

After mentioning some of the moral issues supposedly at stake, the writer concluded, 'In the name of all that is religiously or humanly sacred let the desirability of the whole project be given the closest possible consideration.'

This letter, with its heartfelt, almost pleading tone, rested on very similar feelings to those underlying some of the strongest philosophical arguments which were soon to be raised against laboratory fertilisation. But for the time being, the last word rested with the editor of *Nature*, who returned to the subject two weeks after his first contribution. On 1 March *Nature's* editorial headline asked, 'How sinister is biology?', and commented on coverage of Steptoe and Edwards's work and on John Gurdon's recent experiments on the cloning of frogs.[29] The basic position proposed was that none of the theoretical possibilities raised in popular discussion – for genuine ectogenesis or the production of identical humans – was anywhere near practical realisation, and that 'a good many people seem to be ready to take sides' on these hypothetical questions 'even though the issues have not been defined with enough clarity for them to know what sides there are to choose'.[30]

A few days earlier, *Nature's* editor, John Maddox, had published a very similar piece in *The Times*, in which he asked: 'will the attempts . . . to provide an understanding of and a measure of control over the process of fertilisation also bring a step nearer nightmares such as those of Orwell and Huxley in which genetic engineering is as important as electronic engineering?'[31] Such questions, Maddox reckoned, 'are entirely legitimate and need to be asked', but most of the speculation was unreal because of the limits of present technical capabilities and the severe difficulties of implanting an egg fertilised *in vitro*.

As with the *Guardian* cartoon, the piece embodies the unresolved tension so often seen in commentaries on the science. It is well informed, cool and reassuring, acknowledging fears about future possibilities but arguing that they are a long way off. On the other, the text once again reinforces the association between *in vitro* fertilisation and the popular ideas of what the biological revolution would bring: cloning, test-tube babies and genetic engineering. The brave new world was a long way off, but it was certainly a real possibility.

There the matter rested, so far as public comment was concerned, for almost exactly a year. Then Patrick Steptoe was featured in a BBC television *Horizon* programme, interviewing a woman and apparently agreeing to help her have a child by *in vitro* fertilisation and implantation. This promise of a personal angle transformed the story for the popular press. In 1969 there were no potential parents, no birth in prospect. The only images were photomicrographs from the paper in *Nature*. Now, there was a patient, whose identity (concealed in the programme) was quickly unearthed. The *Daily Mirror*, for example, carried a front-page photograph of the 'Wife waiting for a test-tube baby', Mrs Sylvia Allen, alongside her husband.[32] Now the ambivalence evoked by the work was more acute. The ideology of the family was spelled out loud and clear as warrant for Steptoe's efforts, which were designed to relieve the plight of the infertile wife who believed that 'you don't feel a complete mother until you've had a child of your own. That is what a woman is for.' But along with the personal story came a closer focus on possible hazards of the technique. There was now scarcely any mention of other projected benefits, of improved contraception or the prevention of genetic disease, and several commentators raised the possibility of abnormalities being caused by manipulations of the embryo.

That this shift was due to the personalisation of the story is suggested by the fact that there was little evidence of further scientific advance on Steptoe and Edwards's part. As *Science Journal* pointed out, the original report from their laboratory demonstrated neither fusion of pronuclei nor cleavage of the egg, two crucial stages in the latter part of the process leading to formation of an embryo.[33] On this basis, a recent claim by Dr Douglas Bevis at Sheffield University that he had cultured fertilised eggs through two cell divisions provided more concrete evidence of 'Progress toward [a] test tube baby', as *The Times* headlined its report of the *Horizon* programme.

There was also confusion about how far advanced the Cambridge team were. On the Thursday following the broadcast, Steptoe held a press conference at which he was doubtful about the date of the first attempted implant. 'He said that if any patients were under the impression that it might be possible in the next few weeks or months, there had probably been a misunderstanding.'[34] This contradicted Mrs Allen's expectation, reported in the *Daily Mail*, that the 'final experiment' would take place in two to six weeks. Steptoe also pointed out that the work was being hampered by lack of money, which drew editorial reproof from *The Times* for appealing directly to the public for support, over the heads of the proper authorities.[35]

Whatever Steptoe's motives, there were now stronger critical currents around his work. One was the combined critique of science and technology still being developed by environmentalists. A few weeks before the *Horizon* programme featuring Steptoe, for example, the BBC had launched a new television drama series, *Doomwatch*, which was attracting wide comment. The theme of the series was that 'the honeymoon of science is over',[36] and biomedical threats were often investigated by the *Doomwatch* team, who were generally faced by exactly the 'perverted men in white coats doing nasty experiments' whom Maddox found so irksome. The first programme depicted animal to human heart transplants, and later episodes were to feature genetic engineering in rats, and a scientist secretly keeping aborted foetuses alive in incubators for his experiments. The scientific adviser for the series, Dr Kit Pedler, was quickly asked to comment on Steptoe's latest announcement. His response was to suggest that if the techniques were freely available,

> then you have the means . . . of mass producing people without the advent of a mother at all . . . If you extend the experiment a little bit, it is a question of biological engineering. A general might order 100,000 troops to be produced. This can only be stopped by the public making some sort of objection.[37]

Echoes of these sentiments were heard elsewhere. The *Times* background article the following day was headed 'Move to threshold of genetic engineering' even though the text argued the opposite. A cardinal now supplemented the earlier moral objection from the Catholic Church with the observation that, 'as anyone will realise, the possible effects of developments in long-term "biological engineering" are frightening'.[38] At the end of the week, the *Sunday Times* took up the same themes, in a report of an academic conference in Zurich on 'The Manipulation of Man'. Zoologist Adolf Portman was depicted as one of a series of caricatures alongside the report, saying that, 'with the possibility of breeding human beings, biology is the third World Power'. Italian geneticist Professor Marco Fracerro apparently simply recited the litany. His quoted words, boxed and highlighted, were: 'Manipulation . . . means genes, test-tube babies, Hitler and the Master Race.'[39]

The *Daily Express* appeared to share this view, suggesting in an editorial that, while *in vitro* fertilisation might bring happiness to a childless couple, it would also be a 'major step on the road towards the breeding of totally scientifically produced human beings'.[40] An article on the same page was headed 'Test-Tube babies – soon we'll hardly need people', and was set off by a cartoon which combined comment on the test-tube baby and on a recent BBC sex-education programme. A television announcer told his wide-eyed infant audience – 'Kiddies! Forget last week's lesson. That's out of date and old hat!'[41]

Several newspapers reported a statement by a Birmingham University embryologist which seemed to bear out one of the *Express*'s predictions, of surrogate motherhood. Dr Jack Cohen forecast that women would shortly be able to arrange for others to bear their children, suggesting that £2,000 would be a suitable fee for this service. He went on, blithely: 'Film stars might as a matter of course have their children by host mothers to preserve the body beautiful. For young girls, it would be a lot more attractive than prostitution. There might be professional host mothers. The system is already used commercially in animals.'[42] Cohen's motives for painting this picture are unclear, but it was an early airing of an issue which caused great controversy later in the development of *in vitro* fertilisation technology, and which Steptoe brushed off at the time with a rather terse suggestion that it took no account of female psychology.

This second instalment in the debate about Steptoe and Edwards's work was effectively over a week after the *Horizon* programme was aired, and can be summed up by examining the completely contrasting interpretations of the science correspondents of two British broadsheet Sunday papers which treated the matter at length the following week. In the *Observer*, Gerald Leach attempted to debunk the 'Test tube fantasy'.[43] He had seen a week of 'fevered speculation' and 'mindless escalation into fantasy' which provoked needless concern about artificial wombs, host mothers and mass-produced humans. He suggested that the idea of host mothers was already perfectly feasible, with artificial insemination. The chances of embryo implants being used in this way, in his view, were virtually nil. Each new development should be considered separately, and the kinds of manipulation envisaged as future possibilities were so far distant that they should not be allowed to influence or pass judgement on what was 'simply an artificial aid for a natural process'.[44]

The *Observer*'s rival the *Sunday Times* presented an altogether different view. After relating some of the background to the work, including Edwards's departure from Mill Hill when the Medical Research Council ceased to support his research, the paper's science correspondent turned to its implications, suggesting that 'the fertilisation of a human egg outside the body and its successful implantation in the womb is more than just a treatment for certain kinds of infertility. From it will arise a whole series of developments which raise awkward ethical and practical problems.'[45] These included surrogate motherhood, and attitudes to abnormal children if some were known to be preventable. 'Could it lead to their being regarded not as unfortunate biological accidents, but as administrative errors, for whom someone was to blame?' And there were the social problems likely to occur if sex selection became widely available. All these, he felt, were 'disturbing enough, even without growth to "decanting", as Aldous Huxley called it'.

Unlike his opposite number on the *Observer*, this correspondent felt that there was danger in looking at each new development in a limited medical context, because 'the necessary debate will never take place. Step by beneficial step we may arrive at a place where we do not really want to be.' He concluded with an historical parallel: 'Dr Edwards may care to ponder as he grows his human embryos that opposite his laboratory is the Cavendish where Rutherford in 1934 said that the splitting of the atom would have no practical applications.'

The trailer for this article on the paper's front page ran, accurately enough, 'Test-tube babies: when the scientists have got to stop'. Some newspaper readers certainly agreed, to judge again from the correspondence published by *The Times*. One letter-writer confessed that, 'personally, I find the idea of creating life at man's will terrifying. I can only hope that collectively we have sufficient wisdom to outlaw such scientific endeavour.'[46] Another declared that 'the recent news of the possibility of a "test-tube" baby fills me with despair for the future. I cannot refrain from trying to express my horror and to register some measure of protest.'[47]

So as well as prompting opposing arguments, the possibilities of the work mobilised completely contradictory feelings: hope for some as opposed to despair for others. As with heart transplants, what some saw as a welcome advance, others regarded as unspeakably intrusive. As Leach saw it at the time:

> It is less than three years since the first heart transplants shocked many people, who found them, somehow, an intrusion into – or manipulation of – a terribly deep-seated aspect of what it is to be a human being. Now, it appears to many, the sexual act, another fundamental part of our collective 'mental landscape', has been dragged out and 'manipulated' on the laboratory bench, and therefore degraded in some way.[48]

It was unusual for these more diffuse aspects of the reaction to biological experiment to be addressed explicitly by a journalist, but the feelings energising cartoons about the work suggest this is an accurate diagnosis.[49] And it was similar feelings that motivated the more developed positions on the new reproductive technology which began to appear over the next few years.

Babies and bioethics: 1970–78

Steptoe and Edwards had an unbeatable advantage over their colleagues as their research programme developed over the next seven or eight years. A goal like theirs might have become the subject of a scientific race, as Watson had recently depicted so vividly in *The Double Helix*. But there was no one else in the race. Work along the lines they were pursuing was extremely controversial in the USA, and tacitly proscribed by the Medical Research Council in Britain. They were in the happy position of working in the UK with private American finance, thus avoiding both obstacles.

While they worked on, others debated the possible outcome. In the US, the climate of opinion in the mid-1970s resulted in a government-sanctioned moratorium on foetal research, which, although only officially in force for a year from 1974, effectively halted work in the field until after the birth of Louise Brown – the first baby conceived *in vitro* – in Britain. Such a moratorium was endorsed by several commentators who were prominent in the newly expanding academic area of bioethics.

There had long been a fairly extensive literature on the ethical and legal position of doctors and researchers involved in or with abortion, animal experimentation and euthanasia. In the 1960s, this literature broadened in scope to take in questions prompted by the advance of biomedical science, especially in the ethics of organ donation and transplantation, safeguards for human subjects of research and problems of genetic screening and counselling.[50] The outlines of this new academic discipline were thus beginning to emerge just when Steptoe and Edwards first received wide publicity.

Writers on bioethics who took up the issues raised by *in vitro* fertilisation typically focused on the risks involved for the unborn child after conception. All three of the authors I will look at in detail concluded that any attempt to produce a child in this way must be unethical according to traditional medical standards, because the risks were unknown, and the subject could not give informed consent. On this basis, they wanted the research to stop. Leon Kass, for example, writing in the *New England Journal of Medicine*, called for a 'profession-wide, self-imposed moratorium on attempts to produce human children by means of *in vitro* fertilisation and embryo transfer . . . *at least* until the safety of the procedure can be assessed and assured'.[51]

Mark Lappe made a similar proposal the following year, with the addition of an international review body to oversee the research.[52] Princeton theologian Paul Ramsey went one step further. In an article published in the *Journal of the American Medical Association* in June 1972, he argued that the moratorium should be permanent, since there was no ethically acceptable way of clarifying uncertainty about the risks of embryo research. The work was 'unethical medical experimentation on possible future human beings', because of the undefined risks involved. And, in Ramsey's view, 'we cannot ethically *get to know* how to perfect this technique to relieve humanity of infertility.'[53]

The argument was presented as if this were a philosophical matter, a question of ethics, and the debate was taken up by some writers on those grounds.[54] But a closer look at the work of Ramsey and Kass shows that their attack on *in vitro* fertilisation was motivated by a wider concern about developments in biological science, a concern whose roots lay in the imagery and ideas which had been developing now for several decades.

Kass published his first major article on *in vitro* fertilisation in November 1974. In the same week, a more general paper of his appeared in *Science*, in which he discussed a whole range of developments – technologies for the control of life and death, genetic engineering, and other reproductive manipulations. He began by declaring that 'recent advances in biology and medicine suggest that we may rapidly be acquiring the power to modify and control the capacities and activities of men by direct intervention and manipulation of their bodies and minds.'[55] The new technologies which these advances would offer were unique, Kass argued, because the object on which they would operate was man himself:

> The technologies of energy or food production, of communication, of manufacture, and of motion greatly alter the implements available to man and the conditions in which he uses them. In contrast, the biomedical technology works to change the user himself . . . Indeed, both those who welcome and those who fear the advent of 'human engineering' ground their hopes and fears in the same prospect: *that man can for the first time recreate himself.*[56]

Kass's premise was thus an elaborate paraphrase of the point formulated by Haldane and Huxley forty years earlier, and he made the link explicit by referring to *Brave New World* in his paper, and by paying particular attention to the possibility of ectogenesis. He commented extensively on the potential for 'voluntary self-degradation and dehumanisation' inherent in advances in control over reproduction, and although his discussion cited a wide range of possible developments, the only published research papers he referred to were those from Edwards's laboratory in 1969 and 1970.

In this paper, Kass equated *in vitro* fertilisation with the beginning of

> the transfer of procreation from the home to the laboratory and its coincident transformation into manufacture. Increasing control over the product is purchased by increasing depersonalisation of the process . . . Would not the laboratory production of human beings no longer be *human* procreation?

This argument was quite distinct from the ethical objection based on physical risks which appeared in the *New England Journal of Medicine*, and suggests that the latter was fuelled by a more compelling concern.

Ramsey's case is similar. In the first of his two articles in the *Journal of the American Medical Association* in 1972 he spelled out what was essentially the same ethical argument, but again it was not the sole cause of his concern. He concluded that 'we cannot morally get to know how to perfect [*in vitro* fertilisation] to relieve humanity of infertility (even if, once perfected, it would not be a disastrous further step towards the evil design of manufacturing our posterity)'.[57] In a second paper he developed the second point and, like Kass,

used the image of Huxley's hatcheries to bring home what control over reproduction might mean. Overcoming infertility by *in vitro* fertilisation prefigured
or implied further development of reproductive control, according to Ramsey,
because

> when objection is raised that this should not be done by these means, which
> cannot be guaranteed not to be injurious to the child, the only possible
> answer is and logically must be that those who justify this procedure mean
> in all sorts of other ways to control and predetermine the 'product'.[58]

Later in the same article he returned to *Brave New World*, suggesting that

> it is clear that Huxley discerned something that (without a sea-change of
> values) is self-fulfilling in our culture . . . Even now we can begin to learn
> to be happy with decantation instead of natural gestation, by stifling any
> moral doubts about *in vitro* fertilisation and the implantation of embryos.

So both these bioethicists, writing in the two leading American medical journals, drew on fictional associations of external fertilisation to reinforce a slip-
pery slope argument against the technique. The slope was made to look
steeper by an expansive definition of genetic engineering proposed in a *JAMA*
editorial published with Ramsey's first article, suggesting that

> the popular term genetic engineering might be considered as covering any-
> thing having to do with manipulation of the gametes or the fetus for what-
> ever purpose, from conception other than by sexual union of two persons,
> to treatment of disease *in utero*, to the ultimate manufacture of a human
> being to exact specification . . . Thus the earliest procedure in genetic engi-
> neering might be considered to be artificial insemination.[59]

Kass and Ramsey's position became known to a wider public in the early
1970s. Newspaper articles frequently quoted both men.[60] At least once, at a
symposium in the US, Kass and Ramsey confronted Edwards directly, and
their criticism was backed by James Watson, among others. Edwards replied
that their views were 'ultra-conservative' and 'unacceptable'.[61] Watson's comments followed a widely cited article of his in *Atlantic Monthly*, in which he
suggested that Steptoe and Edwards were paving the way for human cloning.
His position was rather different from the one he adopted in the wake of the

controversy over recombinant DNA research, but the fact that such a celebrated scientist appeared to criticise *in vitro* fertilisation fuelled the debate.
Nor was he the only molecular biologist who came out publicly against the
work. Max Perutz, speaking on Edwards's home ground in Cambridge, was
reported as suggesting that large-scale application of *in vitro* fertilisation could
have horrifying results, 'similar in proportion to the thalidomide tragedy'.[62]

And so the reports, articles and books taking account of these criticisms
proliferated. Published discussion of *in vitro* fertilisation from this point
through to the first birth became highly repetitive, as these new comments
were woven into the fabric of past images, and the whole topic became more
completely bound up with general consideration of a biological revolution.
One more example will suffice to show how, as ever, a skilled writer could
blend the newest material with much older ideas and motifs. The piece in
question, from the *New York Times Magazine* in 1972, was written by Willard
Gaylin, at the time president of the Hudson Institute, one of the two main
centres for bioethics in the US. Like Leon Kass, Gaylin was not averse to
invoking the old fictions to make his point.

His article was ostensibly about human cloning. Gaylin suggested that it
was based on discussions at the Institute. But he gave those discussions a particular jumping-off point, not *Brave New World* this time, but *Frankenstein*.[63]
The title, 'Frankenstein myth becomes a reality', was followed by the secondary theme, in bold type, 'We have the awful knowledge to make exact
copies of human beings'. What this might mean was illustrated by images
bordering the first two pages, small photographs of Mozart and Hitler, in
alternation.

The text proper begins with cloning, using a quote from Watson's article
from the previous year, then turns to *Frankenstein* for context:

> The Frankenstein myth has a viability [*sic*] that transcends its original
> intentions and a relevance beyond its original time. The image of the fright-
> ened scientist, guilt-ridden over his creations, ceased to be theoretical with
> the explosion of the first atomic bomb.[64]

It then briefly sketches the growth of the 'scientific domination of society', as
Frankenstein moved closer to reality. The original was fantasy, but in the twentieth century 'the inconceivable has become conceivable, and . . . we find ourselves, indeed, patching human beings together out of parts'. Cloning is the
best evidence for the advent of Frankensteinian science, exemplified by

oft-cited work by Steward at Cornell with carrot cells, and Gurdon's experiments with frogs. Quotations from Haldane and Robert Sinsheimer suggest that human cloning is a serious possibility, the latter having suggested in 1969 that it was likely 'within ten years'. Only then does the argument broaden to include the less ambitious procedures discussed by Kass and Ramsey.

Unusually, Gaylin makes explicit to some extent the way the various levels of his article are connected. *Frankenstein* is used as a metaphor for the possibility of human cloning. However, as Gaylin suggests, cloning itself is symbolic of a series of other dilemmas raised by biological possibilities. As he put it,

> Cloning commands our attention more because it dramatises the developing issues in bioethics than because of its potential threat to our way of life. Many biologists, ethicists and social scientists see it not as a pressing problem but as a metaphoric device serving to focus attention on identical problems that arise from less dramatic forms of genetic engineering.

Such less dramatic forms, he suggests, might slip into use unnoticed as they evolve in small steps from present-day practice. Artificial insemination and *in vitro* fertilisation are the examples he offers. In this way, the solid fact of Steptoe and Edwards's achievement to date is embedded in a series of layers: of speculation about the prospects for human cloning; of ethical argument concerning the need to consider possible consequences not just of a particular line of research but of other, later developments which might thus be made more acceptable; and the whole wrapped up in the *Frankenstein* myth. The article ends, as it began, with the myth. Now, according to Gaylin, the tragic irony is that *Frankenstein* is no longer a fantasy, 'and that in its realisation we no longer identify with Dr Frankenstein but with his monster'.

So within a short time after the first two bursts of publicity for Edwards and Steptoe a number of serious arguments were put about why it should stop and, especially in the US, these arguments had been put in a way which placed the work in the Frankensteinian tradition. Also in the US, two further factors inhibited the research. One was a concern for regulation of research on human subjects, following disclosures about past practice.[65] This concern was also used by the US anti-abortion lobby following a Supreme Court decision in 1973 to permit abortion of a foetus which was not independently viable.

The abortion controversy became especially heated at this time, and biomedical researchers using human foetal material were ready targets. One *in vitro* fertilisation worker testifying before a congressional committee begged to remain anonymous 'to prevent a deluge of what he characterised as hate mail'.[66] There were allegations of misuse of foetal material obtained by abortionists for medical research, and a successful test case was brought in June 1974 by a Massachusetts Citizens for Life group against researchers testing an anti-syphilis drug on foetuses scheduled for abortion.[67]

In the face of such pressures the Federal Government's deliberations on human experimentation took special note of foetal research. In October 1971 the Department of Health, Education and Welfare issued a statement on human experimentation which declared that 'no implantation of human ova fertilised in the laboratory should be attempted until guidelines are developed concerning the responsibilities of the donor and the recipient parents and of research institutions and personnel'.[68] This was, in effect, a moratorium, and in August 1974, when the bill which created the US National Commission for the Protection of Human Subjects of Biomedical and Behavioral Research became law, the moratorium was made obligatory. It stood for a year, after which research proposals were referred to a national Ethical Advisory Board. However, no further work was sanctioned by this board until after the birth of Louise Brown in 1978.

With this effective prohibition on research in the United States, Steptoe and Edwards's lead was secure. They pursued their project through the first half of the decade, with the aid of a large number of infertile patients, and the implantation which was to result in Louise Brown's birth finally took place in late 1977. It was to be the first authenticated full-term pregnancy after *in vitro* fertilisation, though there were earlier claims to keep the possibility in the public eye.[69] By the time it happened, the whole question of limits to biological inquiry had been aired afresh in the context of regulating recombinant DNA research, also subject to a moratorium for a short while in the US, though a voluntary one this time. Early in 1978, however, Steptoe and Edwards announced their team's establishment of a pregnancy, and the focus of popular attention shifted from gene-splicing and fixed once more on the appealing human story of the real test-tube baby.

The baby of the century

The story of the expected birth of a child conceived after *in vitro* fertilisation immediately attracted widespread and intense interest, as the team involved doubtless expected. The first newspaper disclosure came in April 1978,[70] and was followed by frantic efforts to overcome the elaborate precautions taken to protect the identity of the parents. These quickly succeeded, and the Oldham cottage hospital where the baby was to be delivered was surrounded by an international throng of journalists.[71] Papers vied for exclusive coverage, and the rights were eventually sold to the *Daily Mail* for a reputed £300,000.[72] It became an early example of a story, now very familiar in the British press, in which the build-up to an event was as much concerned with press tactics – conflict between *Daily Mail* security staff and rival reporters; a bomb-scare allegedly started by a foreign journalist in an effort to disrupt Mrs Brown's seclusion – as with the advance of the pregnancy. Others speculated about mother and baby's health. Rupert Murdoch's *Sun* exclusively revealed the baby's sex (wrongly, as it turned out, although it was known to the medical team after amniocentesis). All of this helped to underline that something important was happening.

After nearly three months of this build-up, the child was finally delivered, by Caesarean, just before midnight on 25 July. Now the British tabloid press really let rip. The *Daily Mail* ran ten pages on the story over the next two days, with many photographs. The *Express* ran six pages on one day, even without exclusive access to the parents. There was equally lavish coverage internationally, seven-page cover stories in both *Time* and *Newsweek* in succeeding weeks being the tip of a large iceberg.

There is, then, a wealth of material to help assess public reaction to this historic event. The reaction showed the familiar ambivalence, but the contrasts were more sharply drawn this time than often before. The immediate personal story was widely seen as good news, for the parents and the researchers. In Britain, the delight the popular press took in fostering identification with the Browns was combined with a chauvinistic view of the 'all-British miracle'.[73] Much was made of Steptoe's modest facilities in Oldham, of Edwards's many journeys there from Cambridge. Earlier attacks on their work were now presented as one more obstacle for the determined duo to overcome.[74]

In the US, there were some expressions of resentment from researchers that the British had been able to advance on ground forbidden to them. An unidentified 'expert' suggested to the *Observer*'s New York correspondent that 'this country, the US, has the expertise and the means to be way ahead in fertility research . . . but there is a pervasive unspoken caution holding our researchers back'.[75] Other frustrated American scientists cast doubt on the scientific interest of the work. Malcolm Potts, director of the International Fertility Research Programme, pointed out that fertilisation outside the body was 'something frogs do in a dirty stream'.[76] In the same article, in *Newsweek*, another American criticised Steptoe and Edwards obliquely with a wry suggestion that their work 'should do a lot to help those of us who experiment with monkeys'. The US journal *Science* stayed aloof, with a one-paragraph outline of the bare details and the comment: 'for longer renditions of these facts, see any newspaper or magazine'.[77]

These reflections of nationalistic pride or scientific pique were relatively minor influences on the overall coverage, however. *Nature's* summing up the week after the birth was largely accurate, as far as it went: 'The successful birth . . . of a healthy baby conceived by *in vitro* fertilisation has been almost universally welcomed and hailed as an important advance in the treatment of certain types of infertility.'[78]

The style of the welcome was epitomised by *France-Soir's* headline: 'Victoire de la Science, victoire de l'Amour'. Yet, even so, *Nature's* summary was only part of the truth. For amid all the emotional copy about the 'baby of the century', the accounts of John and Lesley Brown's tearful expressions of gratitude, the photographs of 'Britain's happiest family' together for the first time, there were misgivings. While the achievement of an *in vitro* infant was good news in the present, it mobilised all the familiar fears about the future. Huxley's shadow showed up more clearly than ever. These fears even found their way into the two British papers which took the most emphatically positive view of the birth. The *Daily Mail*, enjoying its exclusive, nevertheless declared that

> even as we share the happiness of John and Lesley Brown . . . we cannot fail to be aware of the unease some genuinely feel.
>
> Amid the rejoicing there are those who shiver involuntarily. 'Where,' they ask, 'is it going to end?' And implied in that question is a chill premonition of a brave new world coming closer.[79]

In the *Daily Express*, right-wing columnist George Gale put the same point more directly. After canvassing the whole range of possibilities in the literature – finishing with the possibility of 'genetic engineering on a massive scale performed by scientists working under the direction of the state' – he suggested how this might come about, by degrees:

> It could very easily happen . . . we already have sperm banks and may soon
> have egg banks. Under the guise of eliminating hereditary diseases and
> defects, governments could easily insist on the 'right' or 'best' kind of
> genetic material to be used . . . Far fetched? Of course it is far fetched. But
> the birth of the Brown baby in Oldham has already fetched us far along
> that road.[80]

His conclusion was blown up and repeated, centre-page: 'If I peer into the
future I see a society of people bred to measure'.

Misgivings like this appeared in one form or another in virtually every
British paper. The *Guardian*'s science correspondent observed that 'Today's
cheers and congratulations, however warm and appropriate, have little to do
with implications.'[81] *The Times* suggested that this was just one of a series of
steps in reproductive biology which raised 'moral questions of great social
importance'. They arose because 'the long-term issue raised by the birth is to
question what this development means for the genetic engineer. It is one thing
to help a childless couple to conceive by overcoming an obstacle to conception . . . It would be another to create new interferences with the processes of
life.'[82]

The interferences most recently in the news had concerned recombinant
DNA techniques, and these were quickly linked with the advent of the test-
tube baby. A House of Commons committee, for example, declined to
endorse proposed British government guidelines on genetic experiments,
which were being considered the day after the birth in Oldham.[83] A government spokeswoman assured the committee that the newly delivered test-tube
baby had nothing to do with genetic manipulation, but the committee apparently found Leo Abse, MP, more convincing when he asserted that women
would soon be able to buy 'frozen foetuses to be implanted into their wombs'.
Abse echoed Hafez's words of some thirteen years before, referring to a time
when 'an embryo purchaser could select in advance the colour of the baby's
eyes and its probable IQ'.[84]

A broader view of the range of opinion in the mainstream media comes
from a more detailed look at the two US news magazines, *Time* and *Newsweek*.
Time's cover story was written before the birth, and appeared on 31 July, and
Newsweek was a week behind. The reports, though, were very similar in
approach. Between them, they incorporated many of the historic images
shaping debate about biology, and the congruence of the interpretative frames
each magazine's team of writers chose is striking.

Compare, for example, the opening paragraphs of each. *Newsweek* began with the baby: 'She was born at 11.47 p.m. with a lusty yell, and it was a cry round the brave new world.'[85] *Time* made a more laboured reference to Huxley, beginning with a lengthy italicised quote from the Director's description of the fertilisation process in the tour of the hatcheries which opens the novel, then suggesting that 'To millions of people in Britain and elsewhere around the world last week, it seemed as if Huxley's prophetic vision had become reality.'[86]

The lead writer for *Time* felt that the story raised 'all sorts of fears we wanted to address'. His account of Steptoe and Edwards's work and of the journalistic furore in Oldham was peppered with further allusions to the historic images, mixed in with quotes from contemporary critics of the research. Frankenstein and Faust's homunculus appeared a few paragraphs before a 'leading obstetrician' voicing his concern that 'The potential for misadventure is unlimited . . . What if we got an otherwise perfectly formed individual that was a cyclops. Who is responsible? The parents? The doctor? Is the government obligated to take care of it . . . ?'[87]

The recombinant DNA debate was also brought in in both accounts. In *Time*:

Some thoughtful observers saw the work as still another step toward further control and manipulation of basic life processes – comparable perhaps to the recently acquired ability of molecular biologists to rearrange and recombine genes of different creatures and even to create new life forms.

Newsweek saw a more direct connection, suggesting that 'recombinant DNA technology might eventually be used to alter the genes of human fetuses just fertilised in the test tube'.[88]

Both magazines also found space for the established views of the critically inclined ethicists. Paul Ramsey's belief that the procedure was unacceptable because of unknown risks to the child, and his fear that test-tube babies would be stigmatised appeared in *Time*. *Newsweek* quoted Ramsey and Kass, warning that, in Ramsey's words, 'The whole idea of designing our descendants, of fabricating the next generation, of making reproduction synonymous with manufacturing, is already in the picture.'

In this fashion, negative views of the research surfaced even when the positive side was being promoted heavily by many commentators. The most laudatory coverage, in Britain, still showed a persistent emphasis on the normality of the birth which betrayed an unease about the procedure. This was partly to negate the usually unvoiced element of the script of this kind of intervention – familiar from all the scientific creations of Hollywood – that a child produced by unnatural means would be marked somehow by the procedure, be misshapen, soulless or monstrous, malevolent or unloved. It was difficult to speak to this possibility directly while Mrs Brown was waiting to have her baby, but fictions of the time had already applied the simple version of the *Frankenstein* script to test-tube babies. At least two recent novels had described the birth of babies which, as the cover of one promised, 'began to transmogrify' after birth.[89]

But the emphasis on normality extended not just to assurances that fears that the procedure might give rise to an increased incidence of birth defects appeared groundless, but also to the many accounts of the Browns as an ordinary, unassuming couple, just like millions of others except for their inability to have children. The normalisation of an extraordinary procedure took in the traditional image of the expectant father waiting for news, detailed descriptions of the parents' reactions immediately after the birth, of the mother's first glimpse of her child, and was reinforced by many photographs of all three together. If *Frankenstein* was a story of failed parenting, the Browns' story would be of the ideal (scientifically assisted) nuclear family.

The way in which the story was constructed around a child who was at the same time unique and perfectly ordinary suggests something of the tension between the contradictory responses to human *in vitro* fertilisation at the moment it became a usable technique. Yes, the procedure held hope for the infertile, who were represented as naturally anxious to overcome obstacles to fulfilling their desire to have normal families. But many nevertheless feared the consequences of the intervention. These fears were rooted less in any academic, social or ethical critique of the work than on an intuition which, if it had been put into words, would have been that no good would come of it. Bringing the whole apparatus of modern biomedical science to bear on the reproductive lives of these two would-be parents was a breach of a boundary which, though long threatened in fiction and speculative writing, had hitherto seemed secure in fact.[90]

Breaching that boundary would bring a whole new set of structures and relations under the sway of the transforming power of modernity, and this was readily seen as a possible consequence of reproductive technologies. As far back as 1969, a few months after Edwards's first paper in *Nature*, a Harris poll which appeared in *Life* suggested that American respondents had a great

fear of infertility, which they saw as a major threat to the stability and permanence of marriage, and this appeared to promote acceptance of techniques
that would reduce infertility. On the other hand, a majority simultaneously
saw new methods of reproduction as a real threat to marriage, 60 per cent
even agreeing that they might mean 'the beginning of the end of family life'.
One commentator suggested that 'children and the couple marriage, the
monogamous family clearly appear in the Harris *Life* poll as *the* central motivation of the majority of Americans today. The more radically a technique
departs from this golden goal, the less acceptable it is.'[91] But where did this
leave a technique which could either reinforce this goal or undermine it?

This was just one aspect of the wider potential of the new biology to
undermine established categories. In the years after Louise Brown's birth,
molecular biology moved from science to technology at high speed. And, just
as *in vitro* fertilisation represented the technological reconstitution of reproduction, the more refined tools of molecular genetics now permit the technological reconstitution of heredity, at a more abstract level. We shall consider
the ways in which we read that story in the closing chapters.

Chapter Nine

The Gene Wars: Regulating Recombinant DNA

> This research is the greatest threat ever to our human existence.
>
> Liebe Cavalieri

> The most overblown thing to enter the American scene since President Kennedy created the fallout shelter.
>
> James Watson

The first test-tube baby was irresistible to news editors around the world because she *was* a baby. The image of a real, embodied human cut through all the abstraction of biology at the molecular level, and showed that the point of the new reproductive technologies was, after all, to produce people. Here was a result of the biological revolution that everyone could immediately relate to.

Nevertheless, debate about biological research was by no means confined to the relatively easy to grasp business of uniting sperm and egg outside their normal meeting place. Indeed, by the time Louise Brown was born, the most technically refined and abstract molecular biology had already given rise to a set of technologies whose introduction was fiercely debated. For historians of biology in public, the years between 1975 and 1980 are the time of the recombinant DNA debate.

The episode has been described in detail by a number of historians and other commentators.[1] Rather like the debate over animal experiments in the 1870s, the significance of the recombinant DNA debate (rDNA for short) lies in the fact that it concerned techniques at the heart of contemporary biological practice, but also served as the occasion for expression of much wider

concerns about the power of experimental biology. In the United States, especially, it exposed concerns about the use of science and technology in the
Vietnam War, about the much-heralded advent of human genetic engineering
and its relation with eugenics, about the possible impact of a wider biological revolution on economy and society, and about the growing commerciali-
sation of science. And it revealed much about scientific attitudes to the public,
as well as public attitudes to science. In this chapter, I will briefly summarise
the course of the debate, and then consider its position in relation to the
longer-term history of public discussion of experimental biology.

Origins

In July 1974 a letter appeared in leading scientific journals which called on
investigators in a new, and rapidly expanding, area of molecular biology to
abstain voluntarily from certain classes of experiment.[2] Its signatories, the
members of a committee formed by the US National Academy of Sciences,
included most of the leading workers in the field. Their self-imposition of a
temporary moratorium on research was widely hailed as a unique and
momentous event in the history of science. It appeared to speak directly to
some of the possibilities discussed by Taylor and others in the 1960s: perhaps
there really was reason to slow research in biology while society discussed the
consequences. But it was not quite like that in practice. The moratorium was
always seen as a short-term measure, and only certain kinds of consequences
were defined as relevant. For the fact that the first discussions which triggered
this episode came from within the scientific community largely determined
its course. In particular it ensured that the ensuing debate was largely
restricted to technical questions about the potential health and environmental risks of the work in question.

The technical questions arose because of the fruitful union of a number of
new findings and techniques, most importantly the discovery of bacterial
enzymes known as restriction enzymes, which cleave DNA molecules in
highly predictable ways, the characterisation of 'plasmids', small circles of free-
floating DNA found in bacteria, and the increasing understanding of genetic
control in these tiny organisms. Together, the new techniques made genetic
manipulation a real possibility. It became clear in the early 1970s that DNA
molecules from different sources – even different species – could be spliced
up and recombined in more or less controlled ways. Genes from a reptile or
a mammal, for example, could be inserted into laboratory-dwelling strains of

the common bacterium *Escherichia coli*, there to replicate and be expressed. This meant that study of the organisation and control of the genomes of higher organisms suddenly became possible. The gene or genes of interest could be abstracted from their normal cellular environment and transplanted into the comparatively simple and well-understood genetic milieu of a bacterium. The techniques promised much in the analysis of the mechanism of gene expression in higher organisms – and in potential commercial application – and were enthusiastically taken up.

But there were also concerns that some proposed DNA transfers might be hazardous. In June 1973, at the US Gordon Research Conference on Nucleic Acids, a presentation on one of the first such experiments provoked discussion of these hazards. The experimenters knew little about the bulk of the DNA in higher organisms, and found it impossible to define the DNA segments inserted into bacteria with complete precision. There was informed speculation about cancer epidemics or the spread of novel antibiotic resistance among pathogens. The chairs of the conference were asked to write to the National Academy of Sciences, suggesting that the Academy recommend 'specific actions or guidelines, should that seem appropriate'.[3] This prompted formation of the committee which in turn wrote the moratorium letter. The committee, chaired by the Stanford University biochemist Paul Berg, concluded that not enough was known about the possible outcome of some genetic manipulation experiments – hence the call for a pause in research. Their open letter called for voluntary suspension of two kinds of experiment which were particularly worrying: those which might confer novel antibiotic resistance or the ability to make toxins on strains of bacteria hitherto without such genes, and those which entailed introduction of genes from animal viruses into bacteria.

They were worried, in other words, about new pathogens, and wanted to address a possible threat to health. As a reporter for *Science* put it at the time, 'the motivation for the Berg group's proposals sprang not from any long-range misgivings about the social impact of genetic engineering, but rather from direct concern about the health hazard'.[4] Indeed, Berg chose to ignore a long letter from Leon Kass, giving a full treatment of wider issues, which Kass, a former molecular biologist himself, sent after they were brought together for dinner by a mutual friend.[5] He believed that the moral issue for scientists was how to protect the public from any immediate dangers from the research. But if that could be guaranteed, there was no problem in seeing the work move ahead.

The signatories to the Berg letter also felt that they were well able to sort out the matter among themselves and their peers. Social responsibility need not lead to any erosion of scientific autonomy, or so they hoped. The scientific community now faced the delicate task of securing self-regulation of these kinds of experiment. After a decade of commentary on the advent of a bio-logical revolution, they had now created a genuinely revolutionary technology. But they really only wanted to discuss how they could proceed to develop it safely, and what precautions they should take against accidental release of organisms with artificial combinations of genes.

By and large, they were remarkably successful. The rough and tumble of debate alarmed many scientists, particularly in the United States, and many came to believe that the regulations they soon had to work under were unduly restrictive. With hindsight, it is apparent that they managed in the end to keep the discussion on their chosen ground. But that was not how it seemed at the time. The suggestion of global hazards, and the call for a moratorium, excited considerable public interest, and for a time a broader discussion looked as though it might overtake the public health debate.

The Berg letter proposed an international scientific conference, which was held at Asilomar, California in February 1975.[6] Here, 140 molecular biologists, with the help of admonitory comments from a few lawyers, hammered out a consensus statement about the conditions under which experiments might proceed. They envisaged that a combination of physical containment (fancy laboratories) and biological containment (specially enfeebled bacteria) could quickly be established which would allow them to do almost all the work which they had been waiting to restart.

This assumption in turn underlay the ensuing regulatory discussion. But after Asilomar there also began a wider questioning of the legitimacy of this line of research. Most conspicuously in the United States, the formulation of policy for containment of hazards – by further conferences, official agencies, national committees and local working parties, and legislators – became caught up for a time in a web of arguments and confrontations on different issues. As the initiators of the moratorium saw it, their action had unleashed a host of unwelcome feelings of hostility directed against their research, and attracted the attention of a variety of public interest groups seeking to exploit the issue for their own ends. Scientists found themselves drawn into discus-sions of the right to free inquiry, accountability and responsibility for the con-sequences of research, the boundary between basic research and application, the right of the public to participate in the assessment of research goals and methods, the nature and likelihood of the possible benefits of the research, the probable social location of their beneficiaries, and the increasing corporate influence on biological inquiry.

Mass psychosis and fear of the unknown

As biologists became more acutely aware of being in a political fix unforeseen by the original whistle-blowers, they developed a range of replies to the calls to take these other issues into account. The scientific defenders of recombinant DNA research had two responses to these expressions of wider concerns. Some doggedly pursued the technical discussion of risk assessment, modes of containment, and regulatory regimes. Others took up the wider issues in a rather general way, but mainly in order to deny their validity. They offered a number of explanations of the strength of feeling which at times surfaced in the discussion. Perhaps the most general was cited by Stanley Cohen, one of the first to use recombinant DNA techniques, and a signatory to the moratorium proposal, who said simply that 'the vague fear of the unknown . . . has been the focal point of this controversy'.[7]

His fellow American Thomas Jukes, writing in *Nature*, suggested that the events in Cambridge, Massachusetts – home of Harvard and MIT – where the city council placed a temporary ban on DNA work within the city limits, were 'somehow typical of a shift in attitudes towards scientific discoveries. Once we were eager to try new things, and none would say us nay. Today, dangers are recognised even before they have been shown to exist.'[8]

This new sensitivity was put on a still broader basis by the Canadian geneticist Louis Siminovitch, who suggested that recombinant DNA researchers were the victim of a 'mass psychosis' whose roots lay in a general reaction to modern life: 'Society has become convinced that everything about us has become overpowering. If we can stop anything, we want to, because we're all thinking, "Don't you think it's time to slow things down?" '[9]

A similar idea was put forward by the Australian physicist-turned-biologist Robert May (today the British government's chief scientific adviser). In a review of the debate in *Science*, May observed that 'the complexities and anomies of the modern world oppress many people, and the recombinant DNA issue is a fine metaphor for these wider ills'. Like others, he connected this kind of feeling with an anti-science movement, suggesting that such ideas could take on 'darker shades of anti-intellectualism'.[10] The Harvard bacteriologist Bernard Davis shared this view, as I observed in my introduction. He

saw the opposition to recombinant DNA work as a manifestation of a generalised attack on science as both a social institution and a way of knowing,
which had now been extended to biomedical research, but had other, more
general causes. He cited four such causes: the perceived consequences of past
technological advances; unfulfilled expectations; the unwelcome impact of
science on philosophical and social ideas; and increasing contact with the
products of science. In similar vein, James Watson argued that molecular
biologists were being cast as scapegoats for all the deeds of science:

> with the recombinant DNA story . . . our opposition is for the most part
> led, not by individuals with any deep knowledge of or even fear of our
> work, but by persons who, for a myriad of reasons, do not like the fruits
> of science, if not of the intellect, and see us as the most vulnerable foe.[11]

The most highly charged of these 'anti-science' accusations came from the
virologist Heinz Fraenkel-Conrat, who recast the debate in terms of a time-
less conflict between progressives bent on the pursuit of truth and fearful reac-
tionaries seeking to obstruct their search for new knowledge. Shortly after
Asilomar, he wrote in a letter to *Nature* that 'The weak, timid, conservative
(not to say conservationist) members of the species may through their
numbers be able to exert real pressure on the Promethean, creative, curious
individuals to make them cease and desist.'[12]

What are we to make of these interpretations of this episode, or of the
equally general view advanced by a number of scholars in science studies that
the debate represented a stage in a renegotiation of a 'social contract' between
science and society?[13] I have been arguing throughout this book that there are
indeed important historical continuities in public responses to science, but I
do not believe that these continuities exist at quite this level of abstraction.
In particular, it is significant that this particular furore was about the latest
developments in *biology* rather than, say, solid-state physics and its contributions to the micro-electronics industry.[14] At one level this was obviously true,
as the regulatory debate was about biohazards. But what of the wider issues
raised by an increasingly powerful experimental biology in the context of
modernity? What evidence is there for echoes of the *Frankenstein* myth in this
round of public discussion of biology?

A close reading of the debate strongly suggests that the answer is that
Frankenstein and his descendants were a presence in the debate, but that most
of those involved actively worked to avoid acknowledging that presence, or

to suppress awareness of it. In that, they were largely, but not completely, successful. They managed to confine debate to the immediate hazard issues associated with manipulating bacteria. But even this debate was framed, at least some of the time, in ways that reinforced the impression of a biology which was attaining powers akin to those envisaged by Mary Shelley.

Thus, alongside the 'official' debate, conducted in committee rooms and in the pages of scientific journals, there were popular accounts of the issues, in which the vocabulary the scientists used had different connotations. Almost everyone, scientists and journalists alike, called the use of recombinant DNA techniques 'genetic engineering'. It was not yet the human genetic engineering which had been widely aired as a possibility since the early 1960s, but the associations were inevitably present. The alternative usage 'genetic manipulation' was scarcely more reassuring. Much the same was true of the word 'cloning', long associated with the idea of reproducing identical organisms, and especially multiple copies of humans *à la* Huxley. Recall Willard Gaylin's *New York Times Magazine* article just two years before the Berg letter. The molecular biologists who now wrote endlessly about 'cloning' pieces of DNA, and debated the merits of alternative 'cloning vectors' were saddled with a technical term which also stood for a possibility which had come to symbolise the powers and responsibilities that might be borne by a biological revolution.[15]

More traditionally, the cutting and splicing of DNA was widely described as life creation, and articles about the health hazards of recombinant bacteria often appeared under headings like 'Creating new forms of life', 'Tinkering with life' or 'People plants'.[16] Some scientific critics of the work also argued at this level, most notably Erwin Chargaff, the by then elderly biochemist whose work on the composition of DNA had provided important data for Watson and Crick. He had waxed scornful about the younger generation of molecular biologists in the past. Now, in rhetoric as charged as Fraenkel-Conrat's he too offered an opinion, that 'Our time is cursed with the necessity for feeble men, masquerading as experts, to make enormously far-reaching decisions. Is there anything more far-reaching than the creation of new forms of life?' His letter, to *Science*, finished with what one contemporary observer called a succinct statement of the deepest fears of all opponents of recombinant DNA work.[17] Chargaff, echoing the environmentalist manifestos of the turn of the decade, was as far from Victor Frankenstein's optimism that 'a new species would bless me as its creator and its source', as Victor himself was after his creation came to life. He offered a conservative view of nature, and prescribed an attitude of stewardship rather than domination:

> This world is given to us on loan. We come and we go; and after a time we leave earth and air and water to others who come after us. Have we the right to counteract the evolutionary wisdom of millions of years, in order to satisfy the ambition and curiosity of a few scientists?[18]

Nor was Chargaff alone among scientists in seeing the advent of recombinant DNA techniques as signalling the start of a process in which one species takes charge of evolution. The American biologist Robert Sinsheimer repeatedly expressed reservations about the direction his science was taking. Not long after Asilomar, for example, in an article in the British weekly *New Scientist* titled 'Troubled dawn for genetic engineering', he was asking, 'Do we want to assume responsibility for life on this planet – to develop new living forms for our own purposes? Shall we take into our hands our own future evolution?'[19] Like Chargaff, though, and like much of the earlier discussion of the biological revolution, this rhetorical questioning was at a very abstract level. But some general press commentaries made the issues less abstract by explicitly identifying recombination of DNA in bacteria with the Frankensteinian tradition. Thus, Arthur Lubow, writing in the US journal *New Times*, began lightly with a suggestion of fairy tales coming true, but then shifted to a more alarming fictional analogy for crossing traditionally conceived species boundaries:

> The mouse who falls in love with an elephant is a likely subject for a droll fable. Everyone knows that animals from different species can't mate . . . But in biological laboratories, modern Dr Frankensteins have found a way to create brand-new forms of life. Perhaps it's a sign of the times that the new creatures are not ten-foot monsters with spokes through their necks, but colorless cells invisible to the naked eye. The DNA of mice and elephants can now be joined in bacteria.[20]

So the *Frankenstein* myth was certainly invoked in some national commentaries. But the shadow of Frankenstein was perhaps most visible in one particular local manifestation of the recombinant DNA debate, in Cambridge, Massachusetts. There many of the elements of the recombinant DNA debate in the USA came together. There were many molecular biologists eager to exploit the new techniques; there were radical scientists, schooled politically during the Vietnam War, who wanted wider public discussion of university research; and there was a local political context which brought the two sides into direct confrontation.

That confrontation took place in the city council chamber, where the populist mayor of Cambridge, Alfred Velluci, called hearings to discuss a proposal to construct a new, high-containment laboratory in the biology department at Harvard. His response to assurances from scientists attending that the recombinant DNA work envisaged for the lab was perfectly safe was to emphasise his responsibility to protect his constituents from 'Frankenstein monsters crawling out of the sewers'.[21] Perhaps he took this framing of the issues from the *Washington Star*, which had recently asked its readers: 'Is Harvard the proper place for Frankenstein tinkering?'[22]

Even though the subsequent course of the controversy in Cambridge yielded little further direct reference to the fictional archetypes, many sensed their continuing power. As in the wider national debate in the US, the scientific effort to contain the discussion was largely successful. The Cambridge Experimentation Review Board (CERB), the first of a number of local citizens' panels set up to consider the safety issues, was a significant novelty in terms of public participation in science policy, but decided at the outset that 'our charge was restricted and that dealing with the public health issues was more than enough work'.[23] Nevertheless, although the board eventually recommended that work could go ahead in Cambridge, they also emphasised that

> the controversy over recombinant DNA research involves profound philosophical issues that extend beyond the scope of our charge. The social and ethical implications of genetic research must receive the broadest possible dialogue in our society. That dialogue should address the issues of whether all knowledge is worth pursuing. It should examine whether any particular route to knowledge threatens to transgress upon our precious human liberties. It should raise the issue of technology assessment in relation to our natural and social ecology.[24]

The popular way of expressing these issues was rather different. Newspaper reporters were constrained by the terms of the CERB's report to write mainly about biohazards and public health. Not so cartoonists. Consider the framing of the board's task offered by the *Boston Globe*'s cartoon commenting on the review board's decision. A mad professor, with 'MIT' sewn on his lab coat, comes bounding through the door, brandishing a newspaper proclaiming that Cambridge Okays Genetic Research. 'Crack out the liquid nitrogen, dumplings . . . we're on our way,' he cries to those within. He is speaking to a weird assortment of misshapen creatures, with a caricature of Karloff's Frankenstein looming above the others. They include a giant ear, a disembodied eye, and a quintet of grotesque hybrids, patched together from parts of different creatures.

As with the cartoons greeting Louise Brown's birth the following year, such a drawing may be read either as an ironic comment on unreasonable fears of what was happening in Cambridge's laboratories, or as expressing those very fears. Either way it suggests an awareness that they had a presence in the debate, and in the review board's deliberations, that such spectres were, in some sense, what the board's remit was really about. But, to repeat, these issues were typically not discussed directly in the context of recombinant DNA, even though they were so widely aired elsewhere. Small wonder that a close observer of the events in Cambridge, the Harvard historian of science Everett Mendelsohn, concludes his essay on 'the public politics of recombinant DNA research'[25] with a strong suggestion that it was the questions not directly addressed which fuelled much of the concern about new experimental techniques. As he wrote:

I believe it is accurate to claim that 'fears' of genetic manipulation were high on the 'hidden agenda' of many of the public critics. Neither the Asilomar gathering nor the city of Cambridge deliberations nor the Congressional efforts provided any entrance into a set of problems that almost certainly stands between science and the public. Modes for internally and externally airing such issues, outside of science fiction, still seem largely out of reach.[26]

Britain: a smoother path for recombinant DNA

So far, I have been reviewing the course of the rDNA debate in the US. In Britain, it is fair to say, the whole business was more muted, for a variety of contingent political and cultural reasons. The tradition of official secrecy helped, of course, as did the lack of a large cohort of young scientists radicalised by the Vietnam War. It was also, perhaps, more compelling to focus on immediate laboratory hazards in a country where there had been recent deaths from laboratory work on smallpox, and where trade unions were anxious to establish a role in framing health and safety policy. In any case, the main historians of the British debate agree with observers of the USA when they comment that, 'in general, diffuse social, ethical and political aspects of the debate were progressively eased out, and the situation came to be defined in the more tractable terms of the technicalities of containment'.[27]

That was certainly the explicit remit of the official British discussions of the matter. The government working party chaired by Lord Ashby echoed their American counterparts by emphasising that 'we were not . . . set up to make ethical judgments about the use of the techniques. Our business has been to assess potential benefits and potential hazards, after discussion with scientists who are familiar with this branch of biology.'[28]

Nevertheless, even in Britain, Bennett and his co-authors think it significant that the period leading up to the Berg letter, Lord Ashby's report and the Asilomar conference was marked by a heightened awareness of possible misuses of genetic technologies. And, they suggest, 'impressionistically, much of the public debate seems to us to have been stylised and stereotyped, drawing its images from "folk memory", and its horrors from *Frankenstein* or *Brave New World*'.[29] In this light, as they say, genetic manipulation was a novel technique which 'breathed new life into long-standing fears'.

> The call for a moratorium, and the subsequent generation of hazard scenarios by some leading scientists, was seen to indicate a threat not only of possible biological pollution by new disease entitities, but also through the potential pollution of the moral environment by scientists 'playing God', altering genetic endowment, and transgressing the natural boundaries between non-interbreeding species.

This observation, using a notion of pollution influenced by the anthropologist Mary Douglas's analysis of taboo and classification systems, is another way of describing the fears of modernity expressed in discussions of experimental biology. And it is especially striking that it appears at the end of a history which concentrates almost exclusively on the course of bureaucratic events in government departments and research organisations in Britain. Like Mendelsohn, these British commentators clearly perceive a 'hidden agenda' in the recombinant DNA episode. It is an agenda which can be made visible in hindsight, by recapturing the efforts some scientific protagonists made to distance the issues they were prepared to discuss from those which they preferred to avoid.

Thus although one cannot divide the reaction to gene-splicing neatly into that due to fears of direct hazards and that evoked by wider implications of scientific advance, the presence of both can be inferred from arguments about exactly what was under discussion. As the British commentators put it, 'what can be said . . . is that as the debate progressed such a distinction became part of the rhetoric of, and actions in, the controversy. There was a broad move by some scientists towards depoliticising the debate by narrowing it to the technicalities of avoiding potential hazards.'[30]

This is only one crucial rhetorical distinction in the debate. As in other episodes in the continuing public discussion about biology, the authority to draw the line between fact and fiction was also a crucial element in the rDNA debate. Hazard scenarios were widely discussed by all parties to the regulatory dispute, and scientists accused opponents of the research of being misled by notions of new epidemics drawn from books like Michael Crichton's *Andromeda Strain*. The concern to establish scientific authority to separate fact from fiction has grown stronger since the 1970s. In the rDNA debate, the main territory for the discussion gave scientists the advantage: they were almost bound to be the most persuasive contributors to a technical discussion of the viability of bacterial strains, the expressibility of genes in different milieux, and the effectiveness of containment measures. But the advantage was short-lived. Framing the issue as a largely technical one bypassed the wider concerns about the technology, but only temporarily.

Chapter Ten

Conclusion: The Human Body Shop

Our stories reflect our own ignorance and fear. They betray the failure of nerve that makes us unwilling to contemplate the future because we cannot imagine a future that is not worse than the present. We are the victims of our self-inflicted cultural degeneration and scientists are our scapegoats.

Allaby, 1995, p. 15

When speculating about the development of new, science-based technologies, participants cannot rely entirely on what they take to be the established facts. While they think and argue about the shape of things to come, they have no alternative but to create some kind of story that goes beyond these facts.

Mulkay, 1996, p. 158

To understand biological function, it is necessary not just to monitor gene expression, but to disrupt and manipulate it. The coming era will require an arsenal of generic reagents for both germline and transient gene disruption.

Lander, 1996

Frankenstein is just a story. It is neither history nor prediction. And for all Mary and Percy Shelley's invocation of contemporary science to suggest that the feat of creation Victor Frankenstein achieves is 'not of impossible occurrence', it does not purport to describe real characters or events.

But the writing of declared fictions is not the only kind of storytelling going on as we deliberate about the paths we will follow in the era of biotechnology and the Human Genome Project. History and prediction, however scholarly, and however carefully built around verifiable facts about past or present, are also kinds of storytelling. In the cultural cacophony of the last years of the twentieth century, there are many contending stories about science, in all these forms. As they are told and retold, they collide and recombine, creating new narratives, sometimes preserving old meanings, sometimes offering new ones. Together, all these stories form part of a diffuse public debate about science and technology, about what research is desirable or permissible, what applications are to be hoped for or feared, about how our society shapes and is shaped by the science it builds. That debate is conducted in many fora, from the primary-coloured media of popular culture to the sober chambers of legislative assemblies. Part of the argument of this book has been that the popular media are as important for reading the state of the debate as the record of official deliberations.

That perception is growing more common, one can observe, as more and more technological projects which once seemed the sole province of fiction appear to bear fruit in reality. The fact that this is itself a commonplace claim in such debates does not mean that it does not contain a large element of truth. This has a number of consequences. One is that the boundaries between fact and fiction are increasingly blurred, according to some, while others seek to enforce a more clear-cut distinction – to devalue 'stories' and restrict discussion to 'facts'.

A second consequence is that there is a greater awareness of the role particular stories may play in debates of this kind. In the case of the *Frankenstein* myth, or the generic 'mad science' script, we recognise more clearly the messages they are most often used to convey. And ours is a time when there is a general increase in cultural awareness of the way such stories work, and in which they become more open to analysis and argument, to humorous or ironic use, to parody or palimpsest, or to outright opposition or subversion. In such a time, when an ironic (self)-consciousness of the role and significance of images in our lives is in the ascendant, and when increasingly complex commentaries – and commentaries on the commentaries – abound, simple readings of recurrent stories are rarely entirely persuasive. Intertextuality is a deliberate feature of the act of composition, especially in film. Every story coexists with alternative narratives, with variants or inversions of the canonical form.

This means that a prospective assessment of the importance of images of

biological science involves interpreting processes which are difficult to summarise neatly, trends which are hard to pin down. But this concluding chapter
is my attempt at such an assessment. First, I will describe some recent fictions
of note which relate to the themes of the book, and which I have not yet discussed. I will then consider their role in substantive debate about current biological research, and especially about scientists' responses to fictional images
of bioscience. Finally, I return to the unfolding of modern biology as a project
of intervening in the body, and consider how that body is now being
conceptualised.

Frankenstein is alive and well and living in Hollywood

By the 1990s, the Frankenstein tradition in the cinema had moved on. Genre
horror successors to the Hammer cycle had virtually ceased, as horror films
largely reverted to supernatural threats, although there was a continuing
interest in television drama based on the story.[1] The noted horror director
Roger Corman released a film version of Brian Aldiss's novel *Frankenstein
Unbound* in 1990, which weaves together the story of the novel with Mary's
life in Geneva, and with a modern science-fiction tale about a time-travelling
inventor from the twenty-first century. However, the Frankenstein theme was
much more often expressed in films about artificial intelligence, cyborgs and
androids.[2]

But just as Hammer saw new cinematic possibilities in the story almost
thirty years after James Whale filmed Karloff, so production values and technology had moved on again and it was, perhaps, only a matter of time before
a major studio was persuaded to back a new version of the tale. Such a version
duly appeared in 1994, directed by Kenneth Branagh, and starring Branagh
himself as Victor and Robert de Niro as his creature.

Billed as *Mary Shelley's Frankenstein*, the result went some way toward realising Branagh's aim 'to use as much of Mary Shelley as had not been seen on
film before'.[3] The film is relatively faithful to the embedded narratives of
the book, retaining Walton's role as another obsessive knowledge-seeker. It
sought to distance itself from the stereotypes of Frankenstein horror by
dispensing with Universal's hunchbacked assistant, and making Victor
Frankenstein 'not a mad scientist but a dangerously sane one'. It highlighted
many of the tensions in the original narrative around parenthood and sexuality, particularly in Branagh's reimagining of the creation scene, at the height
of which Victor and his naked creation stumble around one another, half-

embracing, in a colossal pool of amniotic fluid liberated from the creature's birth chamber.

There were, needless to say, many departures from the text in the interests of realising the director's vision, not least the introduction of an epidemic, with associated forcible vaccinations, in the Ingolstadt streets. And there is an entirely new series of scenes, more Gothic than Mary Shelley ever dreamed of, in which Victor, having once abandoned his effort to create a mate for the monster, reanimates his murdered fiancée by grafting her head on to the body of the executed Justine.

But whether or not the latest major film was faithful to the novel is not of great consequence. The most important point here is that, whatever the differences in detail, the essential elements of the *Frankenstein* myth were once again retained. Branagh saw the question of 'whether brilliant men of science should interfere with matters of life and death' as a contemporary issue. And he brought out both the dual role of the creature as victim and avenger and the ambiguous appeal of Victor as the well-intentioned obsessive.

> This is a sane, cultured, civilized man, one whose ambition, as he sees it, is to be a benefactor of mankind. Predominantly we wanted to depict a man who was trying to do the right thing. We hope audiences today might find parallels with Victor today in some amazing scientist who might be an inch away from curing AIDS or cancer, and needs to make some difficult decisions. Without this kind of investigative bravery, perhaps there wouldn't have been some of the advances we've had in the last hundred years.[4]

At this point in the development of the *Frankenstein* myth, Branagh's testimony is of interest mainly as an example of how one film-maker with a lot of money to spend read the story in the 1990s. No single cultural offering is now going to alter the generally available meaning of the story substantially. The accumulation of past renderings is so great that it outweighs new interpretations of the tale. In any case, the director's own intentions here were mainly in line with the dominant reading of the *Frankenstein* myth. The novelty mainly lay in the cinematic resources brought to bear.

Critical responses were mixed,[5] though in the 1990s none echoed the revulsion expressed by earlier critics confronted with Mary Shelley's text, or Hammer's horror films. But the film, while not reaching box-office heights, did achieve some striking imagery, and its production costs ensured a major promotion budget, itself helping to continue the availability of the *Frankenstein* script.[6] However, while versions of the story which return to the source for inspiration continue to be important, as in the past it is stories which adapt the script to take account of the supposed possibilities of contemporary science that have had the most impact in the 1990s. The other film which does just this, and which demands discussion here, is Steven Spielberg's *Jurassic Park*.[7]

Dinomania

Branagh and his collaborators were drawn to the story of *Frankenstein* because of the contemporary resonances with reproductive and genetic research, but their principal motive was to make a striking film. With the other important product of the tradition of stories about controlling life of the early 1990s, the order of the motives was reversed. At least, that was how Michael Crichton told it in his preface to the novel *Jurassic Park,* published in 1992, and filmed by Steven Spielberg in 1993.

Crichton, who helped invent the genre of the 'techno-thriller' with his first novel published under his own name, *The Andromeda Strain* in 1969, went on to become a well-known popular novelist and film-maker, specialising in topical, often scientific subjects. His career has some striking similarities with that of his fellow American, and fellow medically trained author and film-maker, Robin Cook, though his work is less formulaic than Cook's. In *Jurassic Park*, Crichton introduces his tale with a comment on the advent of an 'unregulated' biotechnology of unprecedented power. When the film was nearing release, he continued to maintain that his intention was 'to provide a serious warning about the dangers of commercialising molecular biology', which he termed 'the most stunning ethical event in the history of science'.[8]

As everyone soon knew, that warning took the form of a story about recreating a variety of dinosaurs from cloned DNA recovered from insects preserved in amber. The plot, which aside from the novelty of dinosaurs is essentially a blend of *Frankenstein* and *The Island of Dr Moreau,* is now familiar to hundreds of millions of people. Universal Studios promoted *Jurassic Park*, with its awesome state of the art special effects, as their summer blockbuster film of 1993, spending a reputed $60 million. They were rewarded with 'a world-wide box-office take of $912 million, the most popular movie of recorded history'.[9] This gives one measure of the distance between the publication of novels in Mary Shelley's day and the mass culture industry of the late twentieth century. Crichton's book, boosted immeasurably by the film, eventually sold over 10 million copies. The 1995 sequel, *The Lost World,* a more or less straight repeat of the first story, was launched with a hardback print-run of two million in the USA, and was immediately translated into twenty-seven other languages. Crichton was already a highly successful author and film director, but the impact of *Jurassic Park* confirmed him as a virtual one-man multimedia industry.

But if the scale and complexity of the enterprise was new, what about the story? As I have said, the plot was a familiar blend, albeit with new elements skilfully incorporated. The main inspiration was to build on the recent scientific speculation that palaeo-DNA might be recovered from amber-trapped insects by imagining that some could conceivably originate from dinosaur blood. The creation of an island theme park populated with these creatures offered an irresistible subject for a film industry which had a long history of using dinosaurs to test the limits of special effects.[10]

Reanimating dinosaurs became another of those ideas which was offered as a symbolic test of the prowess of experimental biology. Palaeo-DNA emerged as a possible area of investigation in the early 1980s, and was of obvious scientific interest for evolutionary studies. But what excited the popular imagination almost immediately was the notion of recreating lost species. And speculation did not, on the whole, centre on even dimly plausible species like the quagga, extinct for around a century, but on dinosaurs, dead for 65 million years.[11] This was in part a product of the long-standing popular fascination with dinosaurs,[12] but also fitted with the pattern of earlier responses to the potential of biological research. As with the idea of the test-tube baby, public images tend to fix on a small number of possibilities to stand for a wider range of outcomes, and to express a reaction to them. By adopting this idea as the premise for his novel, Crichton thus tapped into the much older tradition of speculation about the consequences of experimental biology.

So one can read the constant references to the blood-in-amber scenario in *Jurassic Park,* and in the many commentaries on it which the launch of the film prompted, as a contemporary equivalent of Mary Shelley's claim that the creation on which her story turned 'has been supposed by such distinguished men of science as Humphry Davy as not being beyond the bounds of possibility'. Once again, the stories some scientists were telling were taken up by a storyteller of a rather different kind.

And once again, the author was extremely well versed in current science.

While the basic elements of *Jurassic Park* – a man, an island and a diversity of human-made monsters running amok and reducing all the humans to a beast-like struggle for survival – strike a reader of *The Island of Dr Moreau* as rather familiar, the science and its social relations are bang up to date. Not only is the book packed with capsule explanations of molecular biology, while the film contains an educational animation about DNA, but Crichton also works in regular mentions of chaos theory. In the book, this comes from the mathematician Ian Malcolm, essentially the author's voice, who lectures his companions on our new understanding of the unpredictability of complex phenomena and the fact that it spells the death of the dream of controlling nature. In the film screenplay, Malcolm's contribution is greatly altered and, as Stephen Gould puts it, he is given 'the oldest diatribe, the most hackneyed and predictable staple of every Hollywood monster film since *Frankenstein*: man . . . must not disturb the proper and given course of nature; man must not tinker in God's realm'.[13]

But while the film is more traditional than the book in this respect, some significant new elements do survive the transition to the screen. First, the real villain of the piece is the driving capitalist entrepreneur John Hammond, though he is a less threatening figure in film than book. The scientist who actually puts together dinosaur embryos, the brilliant molecular geneticist Henry Wu, is not mad, merely ambitious. He is seduced by Hammond's promise to free him from the bureaucratic pressures and academic in-fighting of university life.

The heroes of the story, two palaeontologists, are also scientists, albeit of a rather more traditional kind, and critics recognised that its images of science were interestingly mixed. But the closing scene of the book, in which two of the most frightening dinosaurs escape in the direction of the mainland, implies that the aberrant scientific creation of Hammond and his servants cannot be contained. Although the film was of interest to most of the paying millions because of its exploitation of 'dinomania' through stunning new technological special effects, the message about the possible dangers of commercial exploitation of genetic engineering reached an audience lured by the prospect of seeing believable saurians on the screen. The film is a classic, if relatively sophisticated, realisation of the knowledge narrative characterised by Andrew Tudor in his study of earlier cinema, and the knowledge is firmly identified as contemporary science.

Hollywood was not the only venue for production of new variations on the *Frankenstein* myth, merely the most globally visible. Literary-minded producers continued to feel the pull of the story, with notable examples in Britain including Maureen Duffy's *Gorsaga* of 1981, depicting a man–ape hybrid produced by *in vitro* fertilisation. She deliberately inverted what she took to be the roles of creature and creator in *Frankenstein,* making her hybrid warm and handsome and her scientist a cold, unfeeling creature – characterisations which were preserved when the story was filmed by the BBC in 1988 under the title *First Born.* Another novel which made it on to the small screen in Britain was Stephen Gallagher's *Chimera,* filmed by Anglia TV in four parts in 1991. Again, the influence of *in vitro* fertilisation was apparent in the choice of a man–ape hybrid as the creature, and the author suggested that his researches into real genetic engineering had revealed that 'scientists are in a furious race to create a species of powerful sub-humans'.[14]

Some of these authors were trying to voice concerns about science, as well as telling a good story. They were successful, needless to say, when the demands of the story dominated the writing. But reconciling dramatic intent with broader pedagogic or political inclinations did seem to be easier for those who were instinctively uneasy about the new biological technologies. An avowed supporter of biotechnology, the biologist turned science-fiction writer Brian Stableford nevertheless found himself writing a whole book's worth of stories in which the characters face crisis and confusion brought about by new applications of biology. As he explained: 'futuristic fiction is, in the main, much more anxious and alarmist that futurology; this has far more to do with the nature of drama and suspense than with the ideals of authors'.[15]

The literary response to biology, then, remains predominantly negative. How are the scientists responding?

Scientists, anti-science and Frankenstein

Myths only maintain their significance if they are continually retold. The feature of a myth is that the particulars change but the plot remains the same. By that test, the *Frankenstein* myth is alive and well. Its sociological cousin, *Brave New World,* has been incorporated into the vocabulary of debate about bioscience, but it is the *Frankenstein* script which still inspires new stories.

In that sense, my history is one of continuity, of clear lines of descent from Mary Shelley's tale down to the present day. But that thread of continuity has been maintained alongside enormous changes in almost every aspect of life, and especially in the role and powers of science. As these changes have come to be more actively debated and contested, responses to the *Frankenstein* myth have changed as well. Biomedical scientists, in particular, have become very conscious of the way their science may be seen in relation to stories of mad scientists, and of the need to develop political strategies for dealing with this. The discussion of the recombinant DNA debate in the previous chapter brings out clearly how effective these strategies can be. Here, I want to look back at some of the other episodes I have already outlined, to review what these strategies have been.

The simplest has been to warn of the dangers of an 'emotional' response to scientific possibilities. A typical example can be found in *Nature*'s editorial after Louise Brown's birth. The article concluded with a comment on the political implications for science of the possibility of other interventions in the reproductive process:

> On the one hand there may be public pressure to meddle in matters that might be better left alone – in determining the sex of offspring, for example – and, on the other, *there is a danger of an emotional response to such meddling that would backfire on the whole of bio-medical research.*[16]

The writer felt that recent developments in public debate over nuclear energy bore this out. 'The experience of the nuclear industry,' the article continued, 'which attempted to forge ahead without taking public opinion with it has its lessons for biology.'

But how, exactly, to take public opinion along with the science? One common response has been to try and deny the relevance of the issues some claim need to be debated. This can take the form of arguing against the use of popular terms such as 'test-tube babies' or 'genetic engineering'. According to Joshua Lederberg, the latter phrase is as prejudicial as it would be to call surgery 'anatomical manipulation', or scientific nutrition 'moulding a superbaby'.[17]

A similar response is apparent in the way some scientists shift their position as their reading of the political situation changes. Take Philip Abelson. As the biological revolution was becoming established in 1965 he was writing in an editorial in *Science* about how 'now science is producing the basis for . . . great developments, the consequences of which are likely to overshadow those of atomic energy . . . new potentialities which were prophesied by Aldous Huxley shortly after the first atomic explosions'.[18] Six years later, as warnings about the dangers of the biological time bomb were becoming louder, he wrote in the same journal: 'Talk of the dire social implications of laboratory related genetic engineering is premature and unrealistic. It disturbs the public unnecessarily and could lead to harmful restrictions on all scientific research.'[19]

This view, with its illogical suggestion that regulation of genetic engineering was a threat to 'all science', was heard much more widely a few years later when the recombinant DNA debate gave other liberal scientists cause to consider what endorsing the need for public discussion might cost them in terms of research support. Leading biologists who shifted in this wind included Maxine Singer, who supported Marshall Nirenberg's 1967 call for society to prepare for genetic engineering but became a fierce critic of opponents of recombinant DNA research ten years later,[20] and James Watson. In 1971, the future founding head of the National Institutes of Health Human Genome Project published a lengthy article in *Atlantic* expressing strong concern about the possibility of cloning human beings.[21] A few years later, he was suggesting that any such attempt would be of minimal interest.[22] He was also by then a strong critic of the course of the discussions over recombinant DNA.

Denial is one obvious tactic in the face of issues which scientists find uncomfortable. Maintaining privacy is another, exemplified by the attempts to exclude the press from meetings about recombinant DNA research.[23] But there are more sophisticated variants as well. At least three were identifiable by the 1980s. One is to acknowledge that fears about the consequences of experimental biology may be of interest at some future date, but to argue that they are irrelevant to immediate practical concerns. The advances in question are so far distant, and so extraordinarily difficult to achieve, it is suggested, that there is no need to be concerned about them now. This is the position some now adopt in relation to germ-line gene therapy, for example. This tactic plainly has its limitations at a time when the promotional rhetoric used to secure funds for the genome project has reached such heights.[24]

The second is to develop the rhetoric of reason versus emotion, and apply it asymmetrically. As I have argued, the whole course of the recombinant DNA debate exemplifies this. Public discussion was largely restricted to technical questions of risk and hazard. However irksome the outcome for the working researcher, this did confine the issues to matters on which scientific authority was hard to dispute. The concrete hazards were defined in such a way as to seem manageable and others made to seem so nebulous as to be of little importance. The benefits, by contrast, were depicted as both highly desirable and virtually certain. The position was reinforced by the assertion that this was the rational, scientific view, and that critics who saw the issues in different terms were hysterical or irrational. Even June Goodfield's measured early appraisal of the recombinant DNA debate in *Playing God* drew criticism from David Baltimore for 'taking refuge in emotion'.[25]

Finally, the obverse of Abelson's suggestion, that critical discussion of biology threatens all research, is the argument that all consideration of the possible consequences of biological investigation is undercut by the absolute primacy of scientific values. The pursuit of truth is a sufficient good to override all other concerns. A fine example of this argument was the cultural critic George Steiner's Bronowski lecture in Britain in 1978, entitled 'Has truth a future?' Developing his response to the issues raised a decade before in *The Biological Time Bomb*,[26] in the midst of the discussion of recombinant DNA and *in vitro* fertilisation, he argued that humans had no choice but to continue the search for 'truth', even if it did ultimately threaten the society which fostered the search.[27]

So much for the general arguments and tactics that appear when biologists feel the force of public disquiet, and try and defend their right to pursue their research unhampered. But most of this book has been about the influence of fiction – or myth – on framing these debates. How is this aspect of popular discussion treated by the researchers? They are plainly aware of the way biology has been represented in fiction, and the tensions around such images have become stronger as the grounds for the traditional journalistic assertion that 'fiction is becoming reality' have apparently become stronger. While, as I have described, the classic fictions were commonly associated with contemporary science which was used to lend plausibility to their stories,[28] the claim that modern biology provides warrant for stories about controlling life now scarcely needs making. One consequence has been an increasing concern among scientists to establish their authority about what is fact, and what still fiction.

This is a contradictory enterprise. If biologists explicitly contest the reality of particular projects, or the use of particular terms, they may simply reinforce the associations they wish to deny. It is hard to imagine that Lederberg's protestations, for example, however justified they may have been, persuaded any journalists to refrain from writing about 'genetic engineering'. On the other hand, scientists feel that it is difficult to stand by and let what they regard as unjustified claims go unchallenged.

Cloning – facts and fictions

An excellent example of the difficulties this produces is the recent discussion of cloning. By recent, I do not mean simply the furore over the cloned sheep Dolly, who was featured extensively in the world's media in 1997. As I have already illustrated, cloning had long been one of the possibilities used to symbolise the powers of new biological technology. To some extent, this began with Aldous Huxley's vision of Bokanovsky's process in *Brave New World*, but it really emerged as a recurrent motif in debate in the late 1960s and early 1970s. The contention over fact and fiction was then crystallised by a book published by the US science journalist David Rorvik in 1978. In the book, *In His Image*, Rorvik claimed he had assisted in the cloning of a human being. He presented his story as fact, and provided extensive references to back up his claim that such a feat was possible. The claim was strongly denied by scientists, most notably by the British biologist Derek Bromhall, who was angry that his own work had been cited, and provided a convincing demolition of the scientific credibility of the story.[29] Bromhall also accused Rorvik of presenting fiction as fact out of greed, glossing over his own claim about his motives in his epilogue where, some years ahead of Michael Crichton, he expressed the hope that

> many readers will be persuaded of the possibility, perhaps even the probability of what I have described and benefit by this 'preview' of an astonishing development whose time . . . has apparently not yet quite come. And if this book, for whatever reason, increases public interest and participation in decisions related to genetic engineering then I will be more than rewarded for my efforts.[30]

Here, Rorvik seemed to admit that the book was an attempt to exploit the difficulties scientists have in responding to claims like his. By deliberately blurring the boundary between fact and fiction, he hoped to provoke public discussion. In that he succeeded, aided considerably by the welter of scientific denunciation. Although the book was condemned as a literary confidence trick, it was widely read. What was, in truth, a badly written novel, with copious discussion of bioethics culled from the academic literature, became a Literary Guild selection, and the US paperback rights were sold for a quarter of a million dollars.[31]

In addition, Rorvik's earlier non-fiction collection on the biological revolution was republished in paperback,[32] and the controversy spawned numerous newspaper and magazine articles, at least one conventional popular non-fiction book on cloning and a British television documentary.[33] Although this programme generally supported the view that the book was a hoax, the preview article in the *Radio Times*, replete with references to Huxley, concluded that 'in one sense . . . it doesn't matter if it is true or not . . . as the chronology demonstrates, 'if it be not now, yet it will come. The readiness is all'. We *live* in a brave new world.'[34]

With this kind of response, Rorvik seemed justified in believing that his deliberate blurring of genre boundaries would be effective in stimulating debate. Other writers at the time were pursuing the issues highlighted by *The Biological Time Bomb* in fiction, but not getting this kind of attention. Consider, for example, a science fiction novel of the near future by the well-regarded Australian author George Turner, who has a character observe:

> As a group, biologists are the most dangerous men alive. The bomb we've learned to live with and pollution we will handle. But biologists!
>
> What they have achieved since the sixties is enough to put the fear of hellfire into Jehovah himself. Artificial inovulation, the gerontological drugs, brain regrowth and the mechanics of gene manipulation – these are already with us, imperfect and unready but with us.
>
> They are only the beginning.
>
> Consider the implications, and retch.[35]

This is a rather stronger condemnation of contemporary biological research than anything offered by Rorvik, but it is the latter's book which is still remembered, at least when journalists look out their cuttings to write another article on cloning.

The succession of bursts of publicity about cloning show how different parties to the debates about the future implications of biological science treat the role of fictions in these debates in different ways. Some commentators explicitly acknowledge the importance of a fictional tradition in offering symbolic possibilities which dramatise a wider range of concerns about experimental biology. Consider, for example, one notable later media flurry about cloning, when Jerry Hall and colleagues at George Washington University in the US made the front page of the *New York Times* in 1993 with experiments aimed at producing multiple embryos from a single *in vitro* fertilised egg. The lengthy reports which followed in *Time* and *Newsweek* both emphasised the distance between what Hall and colleagues had achieved and the many fictional uses of cloning. Under the heading 'Clone Hype', *Newsweek*'s main report made continual references both to Rorvik's book and to stories about cloned dinosaurs and multiple Hitlers to stress what the work was *not* about. *Time* took a very similar line, while emphasising that the work had created a worldwide sensation, and that 77 per cent of Americans polled that week wanted to see such research either temporarily halted or banned outright. Their reporter observed that the actual research reported 'seems, in many ways, unworthy of the hoopla'.[36] And the magazine made a point of showing how many of the bioethicists who contributed comment to the wider press coverage of the story were offering scenarios that went well beyond existing technology. This suggestion was underlined with a separate article on 'Cloning classics', summarising fictions about the subject from *Brave New World* to Fay Weldon's 1989 novel *The Cloning of Joanna May*. The article opened with the observation that, 'When it comes to dealing with cloning, ethicists and science-fiction writers have almost identical job descriptions.'

The outpouring of concerned commentary which followed Hall's announcement was relatively short-lived. But the most recent episode in the cloning saga evoked a longer-running debate. The cover of *Nature* for 27 February 1997 depicted a Scottish-bred sheep, known as Dolly, over the head-line 'A flock of clones'.[37] By the time *Nature* appeared, the story had already been widely previewed in the general media, following a British Sunday news-paper's exclusive the previous weekend. The birth of Dolly was taken as a signal that cloning of adult mammals, and hence humans, was now a real technological possibility.

A team at the Roslin Institute in Edinburgh, working with colleagues from a local biotechnology firm, had perfected techniques which allowed nuclei from adult cells to be fused with eggs to produce a new individual genetically identical to the original adult. Their success rate was very low, one lamb from 277 fusions, but they did succeed. The enormous volume of comment that followed focused on themes which will mainly be familiar to readers of this book. But there were two particularly prominent features of the discussion which I want to highlight. One was the apparently strongly felt urge, both among media writers and, as opinion polls had already suggested, much of the public, that something must be done to stop human cloning. The other was, as *Nature*'s own editorial argued, that the widespread feeling that governments had been 'caught napping by clones' was an indictment of all the policy-makers, ethical advisers and technology foresight panels who are supposed to be keeping an eye on science and technology. This shows, perhaps, one ultimate limitation of the urge to dismiss particular possibilities which have become the focus of concern about the direction biology may be taking as fictions. The paradoxical result in Dolly's case was that she was the realisation of an idea which had been discussed for more than half a century, yet her advent in the flesh was treated as an enormous surprise. *Newsweek* suggested that

> Twenty years ago, when only the lowly tadpole had been cloned, bioethicists raised the possibility that scientists might some day advance the technology to include human beings as well. They wanted the issue discussed. But scientists assailed the moralists' concerns as alarmist. Let the research go forward, the scientists argued, because cloning human beings would serve no discernible scientific purpose. Now the cloning of humans is within reach, and society as a whole is caught with its ethical pants down.[38]

If so, it was not for want of trying on the part of some commentators. But this time, although there were still respectable arguments why human cloning was unlikely ever to be possible – which some scientists were quick to point out – there was a widely shared determination that the possibility should be considered seriously. Those who responded to Dolly, from President Clinton on down, the former with his call for a report 'within 90 days' from his advisers, wanted to understand more clearly what human cloning might mean.

Although it was widely recognised that cloning was still mainly a symbol for a broader set of technologies, that there were important issues relating to the industrial use of animals connected with Dolly's future as a drug incubator, and that cloned humans would probably not be identical to their genetic forebears, it was the prospect of an end to biological individuality which really caught media audiences' attention. Fictional scenarios abounded as writers tried to help readers think through the possible implications. *Time* magazine, for example, which like *Newsweek* again made cloning its cover story, ran a four-page article on future ethical problems built around a series of tableaux from a hospital cloning laboratory of the future.[39] To top that, it rounded off its special report on Dolly with a tongue-in-cheek science fiction story by Douglas Coupland.[40] The magazine seemed to be taking seriously its suggestion in 1993 that ethicists and science fiction writers had similar jobs.

This seems to suggest that creating fictions about the possible outcomes of applying biological technology to people is a legitimate contribution to debate. Both literary creation and scenario-spinning by bioethicists are ways of alerting society to possibilities which merit discussion before they are realised – certainly a view writers tend to share. Scientists, though, are not so sure. One stance they adopt more commonly is to argue that the public is unable to distinguish fact from fiction. Non-scientists, it is suggested, interpret metaphorical warnings literally. Writers must therefore take responsibility for portraying well-intentioned science in a negative light. Commenting on films in the *Frankenstein* tradition, the distinguished British geneticist Paul Nurse suggests that 'the real dilemma comes when the freedom of the artist to produce what they like has to be combined with the fact that these productions are taken by the public to be an absolutely true portrayal of science'.[41] His British colleague Ruth McKernan agreed that 'the line between science fantasy as entertainment and science fact needs to be drawn more clearly'.[42]

This kind of assertion radically oversimplifies the relations between media and audiences, fact and fiction, and the range of stories available at any one time. 'Factual' stories are always framed in some way which is intended as a guide to interpretation – one reason for the journalistic invocations of *Frankenstein* or *Brave New World*. But this does not mean that the suggestion that some piece of science is 'like' Frankenstein's project or 'reminiscent' of *Brave New World* is taken literally. Both literary criticism and media studies demand that we take a more sophisticated view of what goes on when diverse audiences assimilate a complex set of messages about science.

From contemporary media studies, in particular, we learn that readers, viewers or listeners work actively to construct interpretations of media messages – just as they did in Mary Shelley's time.[43] They are no more likely simply to imbibe the typical message of straight science reporting, characterised by Dorothy Nelkin as 'selling science',[44] than they are to leave Kenneth Branagh's film clamouring for all the laboratories to be closed down.[45] There are always contending interpretations available, if not within a particular text then from other parts of the media landscape, or from the individual context of consumption. Denying this is itself part of the contest over interpretation. As it happens, one of the best examples of the flexibility in interpretation and use of the stories we have been discussing comes from an analysis of scientists' rhetoric in a recent British discussion of embryo research.

Mary Shelley's spectre – Frankenstein in Parliament

The claim that we need a clearer demarcation between fact and fiction has become increasingly common as substantive debates about regulating biological research have become more frequent. A particularly instructive example is the British press and parliamentary debate over embryo research, as analysed by the sociologist Michael Mulkay. This debate followed some years after the birth of Louise Brown, and the publication of a report from a government working party chaired by Lady Warnock. Her report proposed that researchers trying to improve *in vitro* fertilisation techniques should be permitted to experiment with laboratory-created human embryos up to an age limit of fourteen days – held to be the earliest that any precursors of the future nervous system are discernible.

After much delay, the government put before Parliament a draft Human Fertilisation and Embryology Bill which, most unusually, included alternative clauses on research. There was to be a choice between a permissive clause, following Warnock, and a restrictive clause, banning any such work on human embryos, and MPs were given a free vote on the issue. In a series of papers, Mulkay analysed the ensuing debate in detail. Of particular interest are his findings on the role of fictions about biological science in the rhetoric of the debate. His premise is that both sides in such a discussion are telling stories – constructing narratives which go beyond the known facts to predict the future outcome of biological experiments. But what kind of stories do they tell?

As ever, there were one or two invocations of *Frankenstein* in early newspaper reporting of the White Paper (draft legislation). Although these formed a small minority of reports, they reinforced the impression of some scientists that Frankensteinian imagery dominates public discussion of their work. The IVF pioneer Robert Edwards, *Nature* and the *New Scientist* all responded in this way.[46]

However, analysis of later newspaper debate belies this claim. Mulkay read through a sample of eighty-five newspaper and magazine articles written about the embryo bill while it was passing through Parliament. Over one third of them voiced opposition to embryo research, but only *one* of these made any direct reference to science fiction – using Huxley's account of mass production in *Brave New World*. Nor was there much trace of *Frankenstein* or *Brave New World* in the extensive parliamentary debates about the bill. Mulkay finds a few glancing references to the two stories in Hansard in speeches by those opposing research. There was, however, frequent reference to the claimed effect of these stories in speeches by *supporters* of research. Their opponents, they maintained , were unduly influenced by fantastic tales, and this prevented them from grasping the real aims and likely outcomes of embryo research.[47] Those outcomes were represented by the advocates of the fourteen-day clause as essentially benign. Newspaper articles typically featured mothers who had successfully used IVF to avoid the risk of bearing a child with a severe genetic disease. MPs foresaw a wide range of other medical benefits, both in the medium term and in the distant future.

The crucial point, as Mulkay emphasises, is that both sides in the debate are constructing futuristic fictions. Opponents of the research, while largely eschewing mention of *Frankenstein*, were quick to offer speculations about the long-term results of control over reproduction. Supporters, while typically decrying the opposition's futuristic talk as premature, in fact did just the same. But, because there are well-known stories which appear to speak only to neg-ative outcomes, the pro-science lobby was able to apply a double standard. As Mulkay puts it:

> The advocates of embryo research had the advantage of being able to main-tain that their positive vision of the future was supported by the authority of science. Those engaged in opposition to embryo research laboured under the disadvantage that their bleak narratives bore a distinct resemblance to certain familiar stories from science fiction.

For this reason, he argues, while *Frankenstein* clearly offered a rhetorical resource to both sides in the debate, the overall effect was to weaken the oppo-nents' case and strengthen the hand of the embryo researchers. The assertion that anyone with reservations about biological research is taking the old stories literally appears here to have been an extremely effective way of avoiding engagement with the specifics of the critics' case.

New anatomies – stories of the genetic body

Growing a human embryo in the laboratory is an easy concept to grasp. It might begin as a microscopic blob of cells, but it is readily imagined as a miniature human being, open to the lights and instruments of the laboratory. It is thus easy to see how the work that arises from *in vitro* fertilisation is figured as part of the *Frankenstein* tradition in experimental biology.

But what of the cutting edge of contemporary life science, the latest development of the reconceptualisation of life achieved by molecular biology? The abstraction of the new biology, the immersion in a molecular world, the description of the cell and the organism in terms of the 'discourse of gene action'[48] seem worlds away from real, embodied creatures. So, if the *Frankenstein* myth is an expression of fears of bodily intrusion as an implicit goal of the project of modernity, how does it connect with the language of gene mapping and sequencing, of gene probes and gene-splicing?

The connections are many and various, as I have tried to describe. They range from the recurrent suggestion in the early days of molecular genetics that the discovery of the DNA structure amounted to uncovering 'the secret of life' to the persistent habit of cartoonists of depicting the results of genetic manipulation as composite creatures, patched together from the body parts of different species. Recently, though, they have become stronger still, as the idea that the genes are the essence of what makes organisms – including humans – the way they are has been reinforced, both in science and in popular culture. In the latter sphere, as Dorothy Nelkin and Susan Lindee argue, the gene has become a cultural icon, to such an extent that

> Instead of a piece of hereditary information, it has become the key to human relationships and the basis of family cohesion. Instead of a string of purines and pyrimidines, it has become the essence of identity and the source of social difference. Instead of an important molecule, it has become the secular equivalent of the human soul. Narratives of genetic essentialism are omnipresent in popular culture, here explaining evil and predicting destiny, there justifying institutional decisions. They reverberate in public debates about sexuality and race, in court decisions about child custody and criminal responsibility, and in ruminations about the meaning of life.[49]

In science and medicine, the relentless focus on the gene tends toward assumption of the 'genetic body', as described by Gilbert: 'We are what our genes tell us we are. This is a body full of potentials. It is who we are and who we might have been . . . This is the body where the centre of identity is within the nucleus of every somatic cell. It is the body sought by the human genome project.'[50]

So what will be the real impact on medicine, and on the wider culture, of the genome project? There are good grounds for accepting that the change now in prospect is as significant as the genome prophets have said. The molecular biologist and philosopher of science Hans Rheinberger argues convincingly that there is real novelty in the final achievement of an engineering-based understanding of life, and that the novelty lies in the way medicine will now operate on the body.

Medicine has always been a cultural endeavour to maintain the integrity of the body, he writes, 'as a natural entity in its own right and for its own sake'. Evolutionary speculation and eugenics challenged this assumption, but failed to overturn it. Today, in contrast, 'we witness a global irreversible trans-formation of living beings toward deliberately engineered beings, including the human body'.[51] What we are seeing, he suggests, is a transition from representing to intervening. The first technologies of molecular biology were geared to producing 'an extracellular representation of intracellular configurations'. As such, their medical uses were rather limited, as witness the lack of therapeutic progress in sickle-cell anaemia in spite of our intimate knowledge of the protein and DNA lesions underlying the disease. Since the advent of recombinant DNA techniques, there has been a switch to producing 'the intracellular representation of an extracellular project'. That is, that DNA pieces fashioned in the laboratory are inserted in the cell to adapt its behaviour to our own design.

Whether or not this vision turns out to be realisable, it is certainly the kind of idea shaping the agendas of genome researchers. This transformation in ways of thinking about the body is beginning to be understood by the onlookers, outside the laboratories. Again, one can see this in the cartoons. One particular image makes a direct visual link between the advent of Vesalian anatomy and the new anatomies being drawn by the gene-mappers.

As Sawday relates, a common device in early anatomical illustration was the self-anatomising subject, a classically drawn figure, who had parted his (more often, her) chest or abdominal wall to display the organs hidden within. Artists of the 1990s use exactly the same device, but the hitherto con-cealed interior now displays, not a set of dissected organs, but a barcode. The essence of what lies within is not to be thought of at the level of organs, or even cells. It is information. By the same token, there will be a direct link between alteration of this information, by changing the DNA, and altering the body. The genetic engineer, like Frankenstein, is first and foremost an anatomist.

Conclusion

We are never going to be rid of *Frankenstein*, even if we want to be. The story is too deeply embedded in our culture now not to leave its traces or raise echoes whenever we discuss our attitude to science and scientists. And as the products of biological manipulation become ubiquitous, there is every reason for the grip of the story to strengthen. It will remain a powerful symbol of our hopes and fears of a truly effective applied biology's ability both to break down old categories and to offer new ways of shaping our bodies, for good or ill. And it will continue to be evoked, more or less casually, in writing and conversation about biotechnology and the new genetics. As I write, the week's controversy about applied biotechnology happens to be the arrival in Europe of genetically engineered soybeans. Both the British newspaper articles I have to hand, one for the bean, one against, mention *Frankenstein*. Both, though, also argue a detailed case on the basis of a host of technical, social, culinary, aesthetic and regulatory considerations.[52] Once again, the old frame is used metaphorically, to draw attention to a new set of issues in a new context, and the reader can take the reports in a large number of different ways.

But if we have to accept that the old stories will stay alive, it is also surely time to recognise that their use is becoming more limited. If, in the past, they have most often functioned to symbolise or express ultimate consequences, there is now some truth in the suggestion that they hinder rather than help debate when we are confronted with the details of real, novel technologies. This is not because they are fictions which somehow misrepresent the facts. It is because framing debate around *Frankenstein* generally poses all-or-nothing questions which can only be answered 'yes' or 'no'. Mulkay's reading of the embryo debate suggests that this can be used to secure assent as well as rejection, but scientists may find this a dangerous strategy in the longer term.

Frankenstein, simply, tends to polarise a debate which we urgently need to take forward to a point where other answers, more complex than yes or no, are possible.[53] The best, provisional answer to the question, do we want to pursue experimental biology as far and as fast as possible, has to be 'maybe'. It would be unrealistic, as well as undesirable, to call a halt. But nor can we simply continue to build on old ambivalences, simultaneously cultivating dread and desire.[54] We have to find ways of selecting, from the huge ensemble of technologies now on offer, the ones we can feel comfortable about seeing implemented. As ever, the devil is in the detail. And the details of what can really be done with living systems, and how to do it, are now finally becoming available.

The details of what consequences such technologies may have are still to come. Here, we remain in the realms of informed imagination, technically sound speculation, historical analogy – of different kinds of fictions. We will still be looking for ways to articulate our ambivalence about new biotechnological possibilities, possibilities which we may welcome or deplore, or both. And these ways will have to keep pace with a scientific and technological trajectory which is still accelerating.

We will all face the pressures the pace of this science generates – pressures to weigh up new techniques and procedures, to make new decisions, to decide what is acceptable, which paths should be blocked, to make sense of the ultimate promise to reconstitute ourselves. As some of our longest-held hopes and fears for overcoming our bodily limitations come close to fruition, is it too much to ask that we recognise that telling stories about the future is a serious activity, and one that can be encouraged?

These stories can take many forms, and have many meanings. Some narratives are built to motivate their readers to try and realise the ideas they offer, some to warn against other possibilities. If they are tales of the future, no one, scientist or otherwise, can make a final judgement of their veracity. However, we have to grant scientific experts some authority to comment on matters of purely technical feasibility. One problem which seems acute in the current climate is getting them to use this authority constructively. As many of the episodes I have discussed show, there is a continuing political temptation to use technical expertise to close off avenues for debate, not to open them up. This merely perpetuates a mutual distrust between scientists and lay opinion, with researchers accusing the public of taking Frankensteinian fears literally, and laypeople feeling that scientists refuse to respond to their genuine concerns, however knowledgeable and sophisticated they may be.

But one should not be too pessimistic about this. Scientists, after all, live in the world outside the laboratory much of the time, and have as much of a stake in a humane future as anyone else. Many do recognise that the new biology, perhaps more than any previous science, raises questions which merit wider debate. The fact that the National Institutes of Health and the

European Commission have both earmarked modest percentages of their human genome research budgets for ethical, legal and social studies of the new genetics is as much a sign of the times as geneticists complaining that they are being pilloried by a public inflamed by Frankensteinian fantasies.

And perhaps there is at least one earlier precedent for the kind of contribution we can hope for from the people who develop new knowledge on our behalf. One of the most positive features of the recombinant DNA debate was the way in which, at least in the early stages of the discussion, some scientists worked very hard to refine hazard scenarios. They responded to suggestions about possible dangers, not by dismissing them, but by applying their expertise to expand on what the possibilities might really be. True, they focused then on medical and environmental, not social hazards, but the example is encouraging.

It offers, I suggest, one way in which biologists can help us all move beyond our preoccupation with Frankenstein and all his hideous progeny. If they do, it will be because they recognise that scientists can be storytellers, too. And while the old stories may still improve with the telling, the advent of what really is a biological revolution means that we also badly need new stories; many new stories.[55] Some of them, unlike *Frankenstein*, will become real stories. But it is only through telling them that we will enhance our power to choose which ones are enacted in the real world.

Notes

Introduction

1 Rabinow, 1992, p. 236.
2 As she put it in her famous introduction to the third edition of her novel in 1831.
3 Wolpert has used the phrase repeatedly in print. See, for example, Wolpert, 1997.
4 In the United States, for example, the national Science Indicators series sponsored by the National Science Foundation continues to show a high general confidence in the scientific enterprise. The 1996 report, of 1995 data, suggested that three-quarters of the US population believe that the benefits of research outweigh its harmful results. The main author of the report indicated that there was no trace of an anti-science movement in his findings: 'we couldn't find it before, and we can't find it now – and we've looked very hard for it'. See Anon., 'Public faith in science stays high', *Nature*, 381, 30 May 1996, p. 355.
5 Medawar, 1977.
6 Davis, 1991, p. 8.
7 See, for example, Commission of the European Community, 1993; Martin and Tait, 1993. For a general account of the field see Durant, 1992.

8 Compare Shapin's account of science and its publics: Shapin, 1990.
9 Weart, 1988, p. xii.
10 See Farr, 1993. There is ambiguity in the way this idea is treated in the literature. In its original formulation, the idea was concerned with the associations, or 'anchors', evident in the phrases and metaphors a particular group used about a particular subject. As Bauer writes: 'social representations are representations of *something* held by *someone*. It is essential to identify a carrier group, to situate the symbolic content in space and time, and to relate it functionally to a particular inter-group context.' However, in the same paper he also describes how 'a particular representation may change its host and wander among social groups, and take on a life of its own'. It is this which is more likely to be of interest for an account of a longer period, a process more akin, perhaps, to myth-making. *Frankenstein* is perhaps the most striking example of a story which has 'taken on a life of its own' (Bauer, 1994).
11 Schank, 1990.
12 Berman, 1983, p. 15.
13 Ibid., p. 17.
14 Giddens, 1990, p. 1.
15 Neither is this the place to consider the

debates about modernity and post-modernity, except to indicate my general sympathy for Giddens's view that 'rather than entering a period of post-modernity, we are moving into one in which the consequences of modernity are becoming more radicalised and universalised than before'. Giddens, 1990, p. 3.

16 Hall and Gieben, 1992, p. 6.

17 As Robert Romanshyn says: 'Technology is not just a series of events which occurs over there on the other side of the world. It is, on the contrary, the enactment of the human imagination in the world. In building a technological world we create ourselves, and through the events which comprise this world we enact and live out our experiences of awe and wonder, our fantasies of service and of control, our images of exploration and destruction, our dreams of hope and nightmares of despair'. Romanshyn, 1984, p. 10.

18 For example by following Georges Canguilhem's notion of technology as an extension of the body – see his essay on 'Machine and organism', in Canguilhem, 1992.

19 For example, see Barns 1990 and Telotte, 1995.

1   Mary Shelley's Creation

1 Graubard, 1967.

2 For a review, see Cohen, 1966.

3 This was a symbolic recipe rather than an experimental protocol, too vague to follow – a statement of mystical ambition, perhaps. As Vasbinder notes, 'Only an intuitive reading can unlock the secrets of the recipe': Vasbinder, 1984. These figures were one basis for the later stereotyped representation of the alchemist which Roslyn Haynes (1995) identifies as one of the archetypal images of the scientist.

4 Hall, 1969.

5 Quoted by Cohen, 1966, p. 74.

6 As Hall puts it, 'La Mettrie believed that when matter . . . is so arranged as to form the proper sort of fiber, that fiber spontaneously evinces vibrations that are common and fundamental to life processes in general'. Hall, 1969, Vol. 2, p. 55.

7 Quoted, for example, in Toulmin and Goodfield, 1962, p. 315. For a more extensive discussion of the history of automata, and of Descartes and La Mettrie, see also Mazlish, 1993.

8 Mario Praz suggests that Shelley's novel was intended as a direct response to La Mettrie's call for a new Prometheus. See Praz, 'Introductory essay' to *Three Gothic Novels* (Fairclough, 1968).

9 All quotations from Mary Shelley's text are from the Norton edition of the 1818 text, edited by Paul Hunter, 1996; here p. 171. Marilyn Butler (1993) explains in detail why this text is preferable to the revised edition of 1831.

10 Hunter, 1996, p. 172. This kind of reverie was an experience the Romantic writers, who accorded dreams a significance rarely attained before Freud, set great store by. It also helped readers locate Mary Shelley's story firmly in the tradition of the Gothic novel, which was commonly set in a highly stylised dreamscape.

11 p. 34.

12 The combination of power over nature conferred by understanding magnetism and the power to conquer new territories was a common trope in eighteenth-century discussion of natural history, as discussed in detail by Patricia Fara, 1996.

13 See, for example, the critical essays collected in the Norton and Bedford editions of the novel (Smith, 1992, and Hunter, 1996), and in Bloom, 1987.

14 Mellor, 1988.

15 Gilbert and Gubar, 1979.

16 As the great historian Eric Hobsbawm puts it in defining the industrial revolution, there was 'a sudden qualitative and fundamental transformation, which happened in or about the 1780s' in Britain. Hobsbawm, 1969, p. 46.

17 Though Polidori later worked up his effort into a short story, 'The Vampyr', said to be the inspiration for Bram Stoker's *Dracula*.

18 Aldiss, 1975, p. 27.

19 Mary's preface, to her much-revised second edition, was carefully crafted not only to add plausibility (and commercial appeal) to the tale, but also to distance her from the blasphemous cast of the original. But these allusions emphasise the importance of Darwin's pronouncements in shaping the context in which the work was originally placed.

20 King-Hele, 1963; McNeil, 1987.

21 Aldiss, 1975; Scholes and Rabkin, 1977.

22 Reiger, 1974.

23 As Vasbinder notes, in contrast to the vague Paracelsian recipe, 'The artificial man of the novel is described as the product of a clearly defined process which anyone possessing the careful notes Frankenstein made of his experiment could duplicate' (1984, p. 34). This is the most thorough study of scientific sources for the novel.

24 James and Field, 1994.

25 Quoted by Vasbinder, 1984, p. 79.

26 Anon., *Edinburgh Review*, 1803. Quoted ibid., p. 80.

27 Vasbinder concludes that 'all of Victor's scientific studies can be traced in eighteenth-century science. There is not one statement made by Mary that does not have at least one, sometimes dozens, of echoes in the scientific literature of her age'.

28 Following Ruth Richardson's account of the origins of the 1832 Anatomy Act which provided for the bodies of the unclaimed poor to pass to medical schools, Tim Marshall has shown in detail how *Frankenstein* may be read in this context. Richardson, 1988; Marshall, 1995.

29 Quoted in Vasbinder, 1984, p. 81.

30 Sawday, 1995, p. 2. He continues: 'The threat or the reality of violence runs through all Renaissance anatomizations, dissections, partitions and divisions, whether we encounter the term in a medical sense or in a loose metaphorical set of registers. This is not surprising since dissection is an insistence on the partition of something (or someone) which (or who) hitherto possessed their own unique organic integrity.' For another interpretation of the significance of the Vesalian tradition, in some ways comparable, see Romanshyn, 1989, pp. 121–7.

31 In particular, Butler proposes that a dispute between William Lawrence, professor at the Royal College of Surgeons, and his fellow professor John Abernethy led Mary to construct *Frankenstein* as a satire on Abernethy's prescription for studying living beings. Victor fails, in this reading, because he is insufficiently materialist, because he is an *incompetent* scientist. However, Butler concedes that 'even readers of the first edition might have remained unaware of its exploitation of the Abernethy–Lawrence debate; it was masked by so many fleeting references to science at a more popular, less specialized and generally less controversial level'. Butler, 1993, p. xxix.

32 For this last see Adams, 1990, Chapter 6. Mellor also discusses this in the light of Mary and Percy Shelley's own vegetarianism. On the varieties of modern *Frankenstein* criticism see Lipking, 1996, who observes that 'most interpretations serenely interpenetrate, threading through one another like an infinite cat's cradle, and seldom if ever touching'.

33 Mellor, 1988, especially Chapter 5. Mellor's work echoes the historian of science Evelyn Fox Keller's critique of Baconian rhetoric: see especially Keller, 1985. She also argues that Mary Shelley was criticising Davy and the new science, and implicitly endorsing Erasmus Darwin as a more acceptable model of a non-interventionist, evolutionary science: 'Rather than letting organic life-forms evolve slowly over thousands of years according to natural processes of sexual selection [as in Darwin's scheme], Victor Frankenstein wants to originate a new life-form quickly, by chemical means' (p. 100). This seems hard to reconcile with the repeated suggestions, in the two introductions, that Darwin himself had toyed with the idea of life creation.

34 Aldiss, 1975.

2 Hideous Progeny: *Frankenstein* Retold

1 Bram Stoker's novel was much later than *Frankenstein* (1897), but has similar Gothic roots. It is often claimed that the only other story completed from the competition which spawned Mary Shelley's monster, Polidori's *The Vampyr*, was the first printed vampire tale. The later literary, theatrical, cinematic and comic-book history of Dracula has many parallels with that of Frankenstein. See especially Skal, 1990 and Frayling, 1992.

2 Small, 1972, p. 14. Bauer, p. 199.

3 This uneasiness is related to its status as a Gothic novel and not a 'serious' literary production. Even George Levine, co-editor of the landmark collection of *Frankenstein* criticism *The Endurance of Frankenstein* in 1979 betrays signs of this uneasiness, describing the book as a minor novel and one that is deeply flawed by sensationalism, but arguing that it is nonetheless 'the most important minor novel in English'. Levine and Knoepflmacher, 1982, p. 3.

4 Baldick, 1987, p. 1.

5 The most influential theorist of the transition from orality to literacy as a major cultural watershed is Walter Ong. See Ong, 1982. His suggestion that the new visual media, especially television, represent a 'secondary orality', having some of the qualities of older kinds of storytelling, is interesting to consider in relation to *Frankenstein* as myth. But as we shall see, the mythic life of the tale began well before the twentieth century.

6 Baldick, 1987, p. 9.

7 Glut, 1984. This amiably obsessive compendium has a subtitle which captures the intent of Glut's effort: 'Being a Comprehensive Listing of Novels, Translations, Adaptations, Stories, Critical Works, Popular Articles, Series, Fumetti, Verse, Stage Plays, Films, Cartoons, Puppetry, Radio and Television Programs, Comics, Satire and Humor, Spoken and Musical Recordings, Tapes, and Sheet Music Featuring Frankenstein's Monster and/or Descended from Mary Shelley's Novel'. In the unlikely event that anyone wants to bring it up to date, they will presumably have to include holograms, computer software and multimedia presentations, and dance.

8 Jones, 1994.

9 Most recently reprinted in Hunter J., 1996.

10 Martin (1981) gives a convenient summary of what we know about the growth of printed literature, from the Renaissance to the twentieth century. Although the novel was a well-established form by this time, a typical first edition would still only be 750 copies, fewer if library sales were the main expected take-up.

11 As Tropp notes, 'It has been estimated that the reading population of Britain increased from one and a half million in 1780 to between seven and eight million by 1830, a period that roughly coincides with the phenomenal popularity of the Gothic story.' Tropp, 1990, p. 14.

12 Forry, 1990. This useful book reprints the texts of selected dramatisations. See also Lavalley, 1979.

13 For example, 'The New Frankenstein', in *Fraser's Magazine* in 1838. Baldick describes this as 'an unimpressive piece of fictional hack-work, but [it] tells us something of the extent to which the Frankenstein story has been assimilated and accepted . . . as a common imaginative property susceptible to endless experiment and revision. The cliché of the Mad Scientist was taking shape.' Baldick, 1987.

14 See, for example, Hill, 1992. Hill in turn quotes the critic Robert Ray to the effect that the enormous success of early Hollywood cinema was a result of the industry becoming 'intuitively Lévi-Straussian: the American film industry discovered and used the existing body of mythic oppositions provided by the local culture'.

15 Though note here Judith Halberstam's argument that the imaginative response to a text is potentially broader than that evoked by a film. 'One might expect to find that the cinema multiplies the possibilities for monstrosity but in fact, the visual register quickly reaches a limit of visibility. In Frankenstein, the reader can only imagine the dreadful spectacle of the monster and so its monstrosity is limited only by the reader's imagination'. Halberstam, 1995, p. 3.

16 O'Flinn, 1986, p. 203.

17 The film should be seen, but failing that a version of the screenplay is reproduced in Haining, 1994. It also appears in Haining, 1977, where it is credited to

Forrest J. Ackerman, rather than the actual screenwriters, Garrett Ford and Francis Faragoh.

18 For a useful review and reinterpretation, see Grant, 1994.

19 Originally the words were followed by Victor crying out: 'Henry – in the name of God', to which he replies: 'In the name of God, now I know what it feels like to *be* God!' This exchange was cut from the film as released.

20 Hutchings, 1993.

21 Details from Tudor, A., 1989.

22 Ibid., p. 84.

23 Tudor's examples include *The Walking Dead* (1936) for restoring life to the dead with a mechanical heart, *The Man They Could Not Hang* (1939), where a serum achieves the same end, *The Vampire Bat* (1933) which turns on creation of living tissue, as does *Dr X* (1932), and *Murders in the Rue Morgue* (1932) and *Jungle Captive* (1946), for evolutionary meddling. Tudor, A., 1989, p. 140.

24 Schelde, 1993. Schelde takes a folkloric approach to the science fiction film, rather than the horror genre, but his analysis is otherwise very similar to Tudor's. See especially his Chapter 2, 'Dangerous science' and Tudor's Chapter 7, 'Mad science'.

25 Cushing saw Knox as a man who 'had to close his one good eye to the way the bodysnatchers Burke and Hare supplied him with cadavers so that he could show how the human body ticks and thus benefit mankind. For his pains he was hounded out of Edinburgh, just as Frankenstein is always hounded out by villagers'. Cushing, in Haining, 1977, p. 84.

26 All quotations taken from Hutchings (1993), who has inspected 23 reviews of the film in the archives of the British Film Institute.

27 J. Croker (1818), reprinted in Hunter, 1996, p. 190.

28 The year, incidentally, in which Universal Studios returned to the story, not with a film for the cinema but with a three-hour made-for-television version, adapted from the novel by Christopher Isherwood and Don Bachardy, *Frankenstein: The True Story*. I mention it to underline that the screen history of the story includes a number of 'respectable' versions, most often for television, as well as the many horror films.

29 See Hutchings, 1993, Chapter 4, 'Frankenstein and Dracula'.

30 That is, Ken Russell's *Gothic* (1986) and Roger Corman's *Frankenstein Unbound* (1990). The latter is based on Brian Aldiss's novel of the same name.

31 Again, elements of the story may be found in a wider range of films. *Silence of the Lambs* (1991) for example, about the pursuit of a serial killer, portrays a man planning to reassemble a body, or at least its whole skin, from parts stripped from his numerous victims.

32 Further examples may be found in the appendix to Jones, 1994, which discusses TV programmes.

33 Just a taste: a computer search of the text of the British *Financial Times*, perhaps the country's most serious title of all, between 1990 and 1994, yields fifty-two uses of 'Frankenstein'. Aside from references to film and TV and the other arts, the monster was coupled with:

the Channel Tunnel
the poll tax
the US Internal Revenue Service
artificial intelligence
municipal planning officers
a 'monster' recycling plant
Iraq/Saddam Hussein (several times)
privatisation of electricity (that commercial)
a Swedish politician
Soviet central planning
genetic research
the Labour Party
the Department of National Heritage
the Ulster Freedom Fighters
a soccer analyst
'machines' in general
virtual reality
fashion journalism
the (much revived) *New York Post*
US trade laws.

Other newspapers show a similar pattern: a mix of political and technological links, mostly serious, with a scattering of light-hearted uses in other areas.

The pattern with *Dracula*, which has achieved similar currency, is rather different. Of forty-six instances in the *Financial Times* over the same period, only ten occur outside arts reviews or commentary, and most are light-hearted references to effects of the sun or daylight, the colour black (in fashion), or bloodletting. There is also a geographical link with Romania, and specifically Transylvania. The only remotely serious political references are to US Defense programmes (hard to kill) and a reference to Romania's ex-president. There are no direct references to technology.

34 For typical surveys showing the predominance of biomedical coverage see Jones et al., 1978; Einseidel, 1992.

35 In fact, Durant and colleagues suggest, on the basis of findings from a national survey in Britain, that 'medical science may occupy a central and key position within the popular representation of science. In other words, what people know and feel about medicine may help to shape what they know and feel about science as a whole'. Durant et al., 1992.

36 Rostand, 1959, pp. 32–3.

37 Ibid., p. 33.

38 Sussman, 1968.

39 Klingender, 1972; Marx, 1964.

40 Benthall, 1976.

41 Sussman, 1968.

42 Quoted in Berman, 1983, p. 95. Berman's analysis is highly comparable with William Leiss in his account of *The Domination of Nature* (1972), especially pp. 40–4.

43 Berman, 1983, p. 39.

44 Berman mentions Shelley's novel in his discussion of Goethe, suggesting that 'Marx's bourgeois sorcerer descends from Goethe's *Faust*, of course, but also from another literary figure who haunted the imagination of his generation: Mary Shelley's *Frankenstein*. These mythical figures, striving to expand human powers through science and rationality, unleash demonic powers that erupt irrationally, beyond human control, with horrifying results.' But he makes no further mention of *Frankenstein* (ibid., p. 101).

45 Mary Shelley, of course, was well aware of the Faustian frame, which underlies her story, and read and admired Goethe, her contemporary.

46 Formally, the idea that species were defined forms was itself relatively recent, but by the time Mary Shelley wrote it was well established. Thus the great French naturalist Buffon wrote in the 1770s that 'species are the only beings in Nature; Perpetual beings, as ancient and as permanent as Nature herself; each may be considered as a whole, independent of the world, a whole that was counted as one in the creation and that, consequently, is but one unit in Nature'. Quoted in Jacob, 1982, p. 52. In addition, the fact that myth and folklore abounded with monsters, hybrids and chimerae in unknown regions did not mean that the creatures people saw around them did not behave in an orderly way.

47 See Porter, 1991, for a useful commentary.

48 See Rabinow's remark on the Human Genome Project as epitomising modern rationality, as quoted above, p. 2.

3 As Remorseless as Nature: the Rise of Experimental Biology

1 The change in the image lagged somewhat behind scientific practice here. The widespread use of the microscope in the lab is generally dated to the earlier part of the century. According to Bynum: 'although these instruments had been around since the seventeenth century, technical improvements from the late 1820s . . . brought the microscope from the periphery to the centre of medical and biological research'. Bynum, 1994, p. 99.

2 There is a further recent variant on this latest image, as Dorothy Nelkin pointed out to me, that of the molecular biologist as entrepreneur, alongside venture capitalists as a besuited would-be millionaire.

3 See Nelkin and Lindee, 1995 for the history of this image of the gene as a 'cultural icon'. On the genetic body, see Turney and Balmer, in press.

4 Pollack (1994), for example, offers a striking popular account of modern genetic techniques figured as the tools for performing the operations of a 'molecular word processor', operating on the historic collection of texts that is DNA.

5 Secord, 1989.

6 In fact, printing technology underwent a series of incremental improvements from around the 1790s through to the middle of the nineteenth century. This permitted increases like the rise in daily circulation of *The Times* from 10,000 copies in 1820 to 40,000 by 1850, and 60,000 after the Crimean War. See Martin, 1981 for details.

7 Secord (1989) suggests that 'For most readers the creation of life was an experimental fact by the end of January 1837', though his otherwise admirable paper does not offer any direct evidence for this from the readers.

8 Haining offers a racy, though unreliable, account of this history in his book, *The Man who was Frankenstein* (1979).

9 See Coleman, 1971; Fruton, 1972; and Geison, 1978. Coleman, in particular, gives a more detailed account.

10 As Bynum says, Bernard's book is still 'one of the most systematic expositions of a philosophy of science ever produced by a practising scientist'. Bynum, 1994, p. 105.

11 Fleming describes how the disagreements between mechanists and vitalists over the limits of biological inquiry led to a polarisation of positions: 'One group of German mechanists at mid-century were a band of young experimental physiologists who simply wanted to get on with their work, but found themselves intimidated by vitalists who smugly declared it was all in vain – the most fundamental processes of life would always elude a mechanistic analysis. Stung by these taunts . . . a quadrumvirate of rising physiologists – Helmholtz, Ludwig, duBois-Reymond and Brücke – swore a famous mutual oath to account for all bodily processes in physico-chemical terms'. Fleming, 1964, p. ix.

12 On the earlier history of animal experimentation, and of reactions to it, see Maehle and Trohler, 1987.

13 As French puts it, the British antivivisection movement 'was essentially the first of its kind, the intellectual godfather of counterpart agitations in other countries. It was, furthermore, the greatest threat to the existence of experimental medicine in any major country.' French, 1975, p. 12. This fine book is the unsurpassed account of the Victorian history of antivivisection, and my short account here draws on it heavily. See also Bynum, 1994, pp. 168–73.

14 Geison, 1978.

15 Ibid.

16 French, 1975, pp. 229–30. Compare James Turner's account (1980), in which he singles out antivivisection as an expression of the Victorians' new-found sensitivity to pain and suffering.

17 Porter emphasises the strength of the reactions evoked by threats to impose state power directly on the body: 'enforceable smallpox vaccination was briefly introduced into Victorian England, but, meeting fierce opposition, the legislation was watered down; the same is true for compulsory treatment for venereal disease. This solution surely embodies a sense of that inalienable, individual proprietorship of the body stoutly advanced in the secularising formulations of liberal political philosophy from the seventeenth century onwards'. Porter, 1991, p. 216.

18 Turner, 1980, p. 86.

19 As with other episodes in the history of biomedicine, fiction also had a dramatic effect, by increasing awareness of the issues. Bynum suggests that most of the population 'would have been unaware of the issues, or if aware, unmoved either way'. But the controversy did affect images of research reaching much larger numbers than the few thousand antivivisectionist activists. Bynum, 1994, p. 171.

20 Vyvyan, 1971.

21 The same was true to a lesser extent in other countries. Comparable movements elsewhere never posed such a threat to the new discipline. Perhaps prior awareness of vivisection techniques on the continent increased animal-lovers' vigilance in Britain, but

many other factors must be adduced to account for the peculiar appeal of anti-vivisectionism in late Victorian Britain, factors I do not wish to enumerate here. In any case, the overall difference, I suggest, was one of degree rather than kind. Antivivisectionists' efforts secured legislation in a number of other European countries and, somewhat later, promoted regulatory statutes in a number of states of the US. For a wider discussion see Rupke, 1987.

22 Haynes, 1994, p. 109. She goes on: 'Doctors in Victorian literature are seen as heroic figures of unfailing integrity, battling against the diseases of the poor for little financial reward, and the early depictions of natural scientists are endowed with a similar catalogue of virtues, moral and physical as well as intellectual'. Examples include Tom Thurnall in Charles Kingsley's *Two Years Ago* (1857).

23 Burnham, 1971, p. 68.

24 See Chapter 2, p. 37 above.

25 The picture was summarised well by Handlin in a paper from the 1960s which still rings true: 'The linkage of science with practice was clearest, most dramatic and most effective in medicine. This was a field in which ways of knowing were intimately connected with ways of doing. From the mid-nineteenth century onward, the conviction grew that the way to health passed through the medical laboratory . . . By the end of the century the association of hospitals, universities, and laboratories in a firmly articulated complex was the visible manifestation of the union of science and technology'. Handlin, 1972.

26 Turner, 1980, p. 120.

27 Millhauser, 1973. As he says, 'Dr Jekyll is the one nineteenth century fictional character one thinks of immediately, along with Frankenstein; their experiments share the Gothic qualities, being

macabre, grotesque and morally repugnant'.

28 As Martin Tropp points out (1990, pp. 122–3).

29 All three of these American fictions are reprinted in the anthology edited by Franklin (1978), from which my quotations are taken.

30 Millhauser, 1973; emphasis added.

31 Lansbury, 1985. She substantiates her thesis with readings of, for example, Wilkie Collins's *Heart and Science*, together with a range of other Victorian fiction, pornographic and otherwise.

32 The quote comes from the opening paragraph of Ludmilla Jordanova's study *Sexual Visions* (1989), which pursues this theme in much more detail than I can here.

33 Millhauser, 1973.

34 Some experimenters were well aware of this image, of course, as Sir Ronald Ross's tongue-in-cheek short story of 1881, 'The Vivisector', makes clear. It is reprinted in Haining, 1994.

35 Carroll, 1875, cited by French, 1975, p. 303.

36 The language we use to describe observation and experiment also bears the imprint of images established in this period – even when scientists are talking. Paul Rabinow, as visiting ethnographer in the genome labs of the 1990s, reports that 'the obligatory flat joke that greeted me in labs in France and the US was always, "now we will be put under the microscope", or "he's here to treat us like guinea pigs". The lines came from scientists who used neither microscopes (computers and PCR machines) nor employed guinea pigs (yeast and viruses). These jokes disappeared immediately once our work was under way; they reveal an initial anxiety about being objectified, nothing more'. Rabinow, 1996. I agree, but suggest that the way in which the anxiety is expressed is also revealing.

37 Lansbury, 1985, p. 58. Lansbury also describes a celebrated controversy in a later phase of debate – during the Royal Commission on Vivisection of 1907 – which produced uncanny echoes of London's story. Edward Schafer (of whom more in the following chapter) described research at University College London in which he had drowned over fifty dogs: 'He claimed that as the result of thirty-six experiments, two without and thirty-four with anaesthetics, he had discovered a more fruitful method of resuscitating the drowned.' Lansbury records that the popular press 'exploded with indignation': 1985, p. 61.

38 Tropp, 1990, p. 115.

39 See Bergonzi, 1961.

40 Wells, 1895.

41 Wells, 1946, p. 77. All quotations are taken from this Penguin edition. One indicator of the lasting influence of the book is that it was one of the novels reprinted in this uniform paperback edition to mark Wells's 70th birthday. The first printing for each of them in Great Britain was 100,000 copies.

42 Parrinder, 1995, p. 50. Parrinder also reports that in the copy of the book Wells presented to Catherine Robbins, his student and future wife, he drew a sketch of a rabbit dissecting a man.

43 Parrinder, 1972, p. 9.

44 Baldick, 1987, p. 157.

45 *Saturday Review*, 11 April 1896. This review, like the others quoted here, is reprinted in Parrinder, 1972.

46 Raknem, 1962.

47 Hillegas, 1967, p. 37.

48 Bernard, 1865, trans. 1927, p. 103. Although Bernard's classic text was not translated into English until the 1920s, it seems likely that Wells would have encountered the original in Huxley's biology class. The passage continues with a near-echo of Frankenstein's attitude to his work in the charnel house: 'Similarly . . . no anatomist feels

himself in a horrible slaughter house; under the influence of a scientific idea, he delightedly follows a nervous filament through stinking livid flesh, which to any other man would be an object of disgust and horror.' Note that while Victor Frankenstein was simply describing how he felt, Bernard is now making a methodological recommendation.

49 Aldiss, for example, argues that 'the story has become even more topical these days, and certainly no less prone to freeze the blood, because we can take vivisection as a symbol for various scientific procedures available today which are much more effective and radical in their effect upon human or animal flesh: working not merely superficially but at the genetic level. Our powers are infinitely greater than Moreau's; the power to create new animals which will best suit human purposes is already within our grasp. The spirit of Doctor Moreau is alive and well and living in the late twentieth century. These days, Moreau would be state-funded.' Aldiss, preface to Everyman edition of the novel. See also Lehman, 1992.

50 For a review of critical positions, and a discussion of Wells's deliberate adoption of mythic elements, see Bozzetto, 1993.

51 Lehman, 1992.

52 For the full history, see Kevles, 1986.

53 Farrell, 1970, p. 283.

54 Mackenzie, 1976.

55 See Hofstadter, 1959.

56 All, of course, possibilities which became real as the history of eugenics unfolded in the twentieth century in the United States, Germany and elsewhere.

57 Mackenzie, 1976, p. 513.

58 Searle, 1976, p. 113.

59 Farrell, 1970, pp. 288–9.

60 Broks, 1993. Contrast LaFollette's finding, for US popular magazines later

in the century, where she claims there is little difference between the (generally positive) non-fictional images and a much smaller number of fictional depictions of scientists. LaFollette, 1990.

61 Haller, 1963, p. 59.

62 Sources on science fiction in this period include Clareson, 1959, an annotated checklist of American science fiction from 1880 to 1915, and Moskowitz, 1973. See also Evans and Evans, 1971 and Franklin, 1978.

63 Recently reprinted in Haining, 1994.

64 See also Clute and Nichols, 1993, p. 541.

65 See ibid. p. 727.

66 See Mullen, 1967, on which the account which follows draws heavily.

67 Ibid.

68 Mullen maintains that the model for Lazaroff was the Russian-born biologist Metchnikoff, whose theories that medicine would soon enable man to live for 150 years had been discussed at length in an essay of Wells. See Wells, 1914.

4   Creating Life in the Laboratory

1 Williams, 1900, p. 774.

2 Huxley, T.H., 1869, p. 129.

3 Geison, 1969, p. 273.

4 Gruenberg, 1911.

5 Goodell, 1977.

6 Tobey, 1971.

7 Full details are in Pauly's illuminating biography: Pauly, 1987.

8 Jane Maienschein (1991) recounts the intimate relations between these three and a fourth leading American biologist, E.G. Conklin.

9 As François Jacob puts it: 'With the sea-urchin egg, the study of the cell and of embryonic development ceased to be purely observational, and became experimental.' Jacob, 1982, p. 213.

10 See Pauly, 1987, Chapter 5.

11 Robertson, 1926, p. 114.

12 Reingold, 1962. Loeb's first paper reporting artificial parthenogenesis is 'On the nature of the process of fertilization and the artificial production of normal larvae (plutei) from the unfertilized eggs of the sea urchin' (Loeb, 1899). More widely noticed was 'On artificial parthenogenesis in sea-urchins' (Loeb, 1900). Note the shift in terminology.

13 Pauly, 1987, p. 100.

14 As inspection of the *New York Times* index and the *Reader's Guide to Periodical Literature* reveals.

15 Anon., 1911, 'Life still hides its secret', *New York Times*, editorial, 9 August.

16 Anonymous editorial, *New York Times*, 3 December 1913.

17 Tobey, 1971, p. 64. Note, though, that Pauly quotes Wilson writing in the *International Monthly* in 1900 that Loeb's experiments opened up possibilities for 'creating wholly new organic forms by varying slightly the conditions of development'.

18 See De Kruif, 1923; Osterhout, 1928; and Duffus, 1924.

19 Roux, 1908.

20 Tobey, 1971, p. xiii.

21 Dubos, 1976, p. 39.

22 Loeb, 1964.

23 Snyder, 1912. On relations between Snyder, a widely published science writer of the period, and Loeb, see Pauly, 1987, pp. 102–3.

24 Loeb, 1964, p. 14.

25 Anon., *San Francisco Choronicle*, 28 February 1905.

26 Anon., *New York Times*, 26 September 1907.

27 Serviss, 1905.

28 For a full biography of Carrel see Edwards and Edwards, 1974.

29 Corner, 1964.

30 Ibid., p. 234.

31 Harrison's original paper is 'The growth of living nerve cells in vitro' (Harrison, 1907). Passages from this paper are

reprinted in Rock, 1964, pp. 159–63.

32 Witkowski, 1979, p. 279.

33 Edwards and Edwards, 1974. Present-day knowledge indicates that the 'immortal' chick cultures were not in fact possible. Witkowski discusses the various possible explanations for the apparent immortality of Carrel's culture and concludes that there is evidence that at least one of Carrel's technicians periodically refreshed the culture with new cells, to conform with her boss's expectation. Witkowski, 1980.

34 This and other quotations are taken from items in the files of the Rockefeller University Library, New York, formerly the Rockefeller Institute. They are denoted subsequently thus (Rockefeller Collection). *New York Times*, 26 October 1912 (Rockefeller Collection).

35 Anon. (1911) 'Sees way to control growth of tissues', *New York Times*, 21 May.

36 Carrel, 1948, p. 107. First published in English in 1935.

37 Corner, 1964, p. 163.

38 Anon. (1912) 'Close to mystery of life, says Loeb', *New York Times*, 21 July.

39 Hendrick, 1913.

40 Anon., editorial, *The Nation*, 74 (16 January 1902), p. 46.

41 Anon., editorial, *New York Times*, 26 September 1907.

42 Unidentified cutting, 6 November 1912 (Rockefeller Collection).

43 Anon., editorial, 'Life and current science', *Independent*, 54, (1912), p. 2256.

44 Stockridge, 1912; original emphasis.

45 At one stage, Flexner, the Institute director, announced the appointment of a special assistant to take care of Carrel's dogs. See Corner, 1964, pp. 85–7.

46 *The Light*, San Antonio, Texas, 17 November 1912 (Rockefeller Collection).

47 Corner, 1964, p. 59.

48 Anon., 'Head of Pasteur Institute proposes to replace electric chair by laboratory', *New York Sun*, 8 December 1912 (Rockefeller Collection). On the history of medical dissection and popular fears see Richardson, 1988.

49 Anon., *New York World*, 26 October 1912 (Rockefeller Collection).

50 Anon., *Buffalo News*, 27 October 1912 (ibid.).

51 Anon., *New York Press*, 30 October 1912 (ibid.).

52 Anon., *Buffalo Express*, 6 November 1912 (ibid.).

53 The story in question is called 'The Man Who Masters Time', by Cummings, cited in Florescu, 1977, p. 185. I have not so far been able to trace it further. It may be that Florescu has in mind 'The Man Who Made a Man', by H.O. Cummins, first published in *McClure's Magazine* in December 1901. It is reprinted in Haining, 1994. Other stories in this collection underline the fact that the Frankenstein tradition was very much alive in popular literature in this period.

54 As Alvar Ellegard wrote about the 1860s, and as is still partly true even today: 'the BA meetings were used as occasions for a stocktaking of scientific progress during the year. The editorial comments on the meetings – and especially on the presidential addresses – provide us in fact with almost the only information we have on the opinions of an important and influential section of the public – the newspaper editors – in regard to science. To these men, and to that large majority of their readers who did not read scientific periodicals, the BA meeting was the scientific event of the year.' Ellegard, 1958, p. 65.

55 Livingston, 1927.

56 Geison, 1978.

57 Lankester, 1913, p. 288.

58 The address was widely published in

scientific and medical journals as well as in the press. The most accessible is *The Lancet*, 7 September 1912, pp. 676–85.

59 Anon., 'Scientists and the secret of life', *Daily Mirror*, 5 September 1912.

60 Anon. leader, 'Origin of life', *Guardian*, 5 September 1912.

61 These are preserved in the library of the Wellcome Institute for the History of Medicine.

62 Lankester, 1913.

63 Jolly, 1974 is the most accessible biography of Lodge.

64 Anon., 'Creation by chemistry', *Daily Mirror*, 6 September 1912.

65 Anon., leader, 'Synthetic rubberism', *Daily Mail*, 6 September 1912.

66 Quoted in Anon., *Public Opinion*, 102, (1912), p. 241.

67 Ibid.

68 Anon., *Yorkshire Post*, 5 September 1912. Despite this, the writer went on to say that the things created would not really be living organisms on Schafer's scheme, but 'chemo-physical automata'.

69 Anon., 'Origin of life', *Guardian*, 11 September 1912.

70 *The Lancet*, 7 September 1912, p. 703.

71 Anon., leader, 'Reality and physical science', *The Times*, 13 September 1912.

72 Anon., editorial, 'Can life be created? Laboratory-made human beings. Our view', *Evening News*, 5 September 1912.

73 Anon., editorial, 'More life', *Daily Mirror*, 6 September 1912.

74 Untitled letter, *Daily Mirror*, 9 September 1912.

75 Gibbs, 1912.

76 Anon., editorial, 'Life as per recipe', *Daily Herald*, 6 September 1912.

77 Bud, 1993.

5   Into the Brave New World

1 Anon., *New York Times*, 5 January 1936.

2 Hobsbawm, 1994.

3 See Rose and Rose, 1970; Poole and Andrews, 1972; and especially Austoker and Bryder, 1989. On the organisation of research more generally, see Cardwell, 1972.

4 Kohler, 1978.

5 Allen, 1978, especially pp. 73–113. Allen offers an extended discussion of the role of wider social influences in this shift.

6 De Kruif, 1923.

7 Wells et al., 1930, Vol. 2, p. 321; emphasis added. For the background to the project, conceived as a money-making successor to Wells's *Outline of History*, see Huxley, J., 1970, Chapter 7.

8 Wilson, 1923.

9 Wells et al., 1930, Vol. 2, pp. 434–5.

10 Anon., 'Prophecy that scientists will create life', *Daily Mirror*, 12 September 1928, p. 3.

11 Anon., 'Scientist protests', *Daily Mirror*, 14 September 1928.

12 Anon., 'Professor Hill's task', *Daily Mirror*, 15 September 1928.

13 See, for example, Fournier, 1928.

14 See Chapter 4, p. 66.

15 Anon., *Scientific American Monthly*, March 1921, p. 221. Stevenson, 1929.

16 Shipley, 1931.

17 Kaempffert, 1935.

18 Waddington, 1969.

19 Clarke, 1979, p. 227.

20 Kaempffert, 1936. Introduction.

21 Clarke, 1979.

22 Carter, 1977, p. 203; original emphasis. This passage continues: 'not until another century of industrial progress had gone by would the next logical step in fiction be taken, that of applying to the creation of artificial human beings the same techniques of mass production that were liberating and degrading real people'.

23 Quotations from *R.U.R.* are from the Oxford University Press edition, 1962.

24 Čapek, 1923. In this piece, written to set straight what he saw as incorrect interpretations, Čapek declared that 'The

old inventor, Mr Rossum, is . . . a typical representative of the scientific materialism of the last century. His desire to create an artificial man – in the chemical and biological, not the mechanical sense – is inspired by a foolish and obstinate wish to prove God to be unnecessary and absurd. Young Rossum is the modern scientist untroubled by metaphysical ideas; scientific experiment to him is the road to industrial production. He is not concerned to prove but to manufacture. To create a Homunculus is a medieval idea; to bring it into line with the present century this creation must be undertaken on the principle of mass production.'

25 Harkins, 1962, pp. 84–5.

26 In this respect, *R.U.R.* bears comparison with Fritz Lang's great silent film *Metropolis* of 1926. For a detailed analysis of the film, including its relations with the *Frankenstein* myth, see Chapter 6 of Jordanova, 1989.

27 Čapek himself protested at the later use of 'robot' to denote a machine, insisting that he deliberately depicted *his* robots as organic, biological creations. See 'The author of the robots defends himself', *Science Fiction Studies*, 23 (1996), pp. 143–4.

28 Squier, 1994.

29 The essay's continuing appeal is underlined by its recent republication, with a series of contemporary commentaries by distinguished authors which are nevertheless considerably less interesting than Haldane's own text: Dronamraju, 1995. Dronamraju's introduction gives useful background on the original. For further biographical material on Haldane see: Clarke, 1968a and *Dictionary of Scientific Biography*, Vol. 6, pp. 21–3; Pirie, 1966; Dronamraju, 1969; Werskey, 1971, 1979. On the *Today and Tomorrow* series see Stableford, B. (1985), pp. 153–4. For a

different view of this period, and this group, see Robert Bud's history of biotechnology, Bud, 1993, especially pp. 63–6, 71–9.

30 Ibid.

31 Ibid., p. 50.

32 Ibid., p. 40.

33 Clarke, 1968, p. 70.

34 Haldane, 1924, p. 64.

35 Ibid., pp. 65–6.

36 Clarke, 1968, p. 73.

37 Note that the old-Etonian aristocrat Haldane was not yet an avowed communist, as he was in the 1930s and 1940s, but was moving toward radical politics. See Werskey, 1979. Haldane, who remained childless through two marriages, continued to hold an optimistic view of ectogenesis to the end of his life. His posthumously published utopian novel *The Man with Two Memories* (1976) also depicts ectogenesis.

38 Russell, 1924.

39 Russell, 1931, p. 169. Thanks to Hayley Jarvis for drawing my attention to this portion of the book.

40 This creature normally spends its whole life in a non-amphibious form, and Huxley acted on a conjecture that it could nevertheless be induced to change into an 'adult' form, as related species do.

41 The *Daily Mail*, as described in Huxley, J., 1970, p. 126. See also Squier, pp. 36–7. On the general background to Huxley's popularisation, see Kevles, Lemahieu and Patten's chapters in Waters and Van Heblen, 1992.

42 Huxley, J., 1937, p. 92.

43 Ibid.

44 Ibid., p. 33; emphasis added. Note that this was written shortly before Müller successfully induced artificial mutations in *Drosophila* with X-rays, in 1927.

45 One further example: his essay on 'The determination of sex' concludes with the assertion that 'the knowledge we

have acquired of the sex chromosomes is bound in the not-too-distant future to lead to a considerable measure of control over what until recently was one of the greatest mysteries of life'. Ibid., p. 54.

46 Huxley, J., 1926. The story appears in a number of anthologies, for example Conklin, 1965, pp. 146–70.

47 Original emphasis. In the *Cornhill Magazine* version, these remarks were addressed to the 'great British public'; in the *Yale Review*, the national adjective was omitted. There are a number of other minor differences between the two versions.

48 For the history of the magazine, see Carter, 1977.

49 Stableford, 1979.

50 Sapiro, 1964.

51 Roslynn Haynes remarks that 'expressions of fear and mistrust in relation to biologists' are also relatively common in the mainstream fiction of this period, in stories like Maurice Renard's 'Le Docteur Lerne' or Somerset Maugham's 'The Magician'. Haynes, 1994, p. 202.

52 Kline, 1926.

53 Harris, 1929.

54 Kleier, 1928. The disembodied head or brain is a common motif in these stories. See, for example, 'The Living Test Tube' (Simmons, 1928).

55 Snyder, 1926. Another example of an explicitly anti-reductionist story is 'The Plague of the Living Dead' (Verrill, 1927).

56 Flagg, 1927.

57 Ibid.

58 Note that the tales reviewed here all pre-date the cinematic rebirth of *Frankenstein* in 1931, discussed on pp. 29–32. In addition, the science fiction pulps often filled space by reprinting the classics. 'Blasphemer's Plateau', for example, discussed above, appears in *Amazing Stories* immediately following

an episode in a serial reprinting of *The Island of Dr Moreau*.

59 Burroughs, 1929. See also his 'Synthetic Men of Mars', discussed in some detail by J.P. Telotte, 1995, pp. 41–3.

60 See below, Chapter 6, p. 129.

61 Schachner, 1934.

62 Hamilton, 1936. The collection of naturally occurring *Drosophila* mutants was of course well established as the mainspring of experimental genetics in the laboratory of Müller's first mentor, T.H. Morgan at Columbia.

63 The added resonance here is the prevalence of eugenically inspired stories in this period which were concerned with fitness for reproduction. See Nelkin and Lindee's account (1995), especially Chapter 2. Alascia's daughter reacts with understandable revulsion, but can also be read as taking eugenic advice to heart. Müller himself was a lifelong eugenist, as discussed below, pp. 128–31.

64 Though an exceptional one in some ways. Helen Parker suggests that 'until about the middle of this century, genetic stories were generally about mutations in the non-human world, radically mutated plants or animals that threatened the human community and had to be eliminated. Only a few early writers dealt with human mutation.' Parker, 1984, p. 13.

65 Clarke, 1979, p. 240. On Birkenhead (F.E. Smith before he was ennobled), see *DNB*, 1922–30, pp. 783–9.

66 Haldane, ever ready for a fight, especially with a peer of the realm, accused Birkenhead of plagiarism; this led to a spirited exchange in print in which he established the overlap between Birkenhead's text and his own in rather convincing detail.

67 Birkenhead, 1930.

68 Ibid., p. 178.

69 Ibid., p. 61.

70 Ibid.

71 Huxley, 1950, p. 9.

72 In fact he had been toying with the idea of ectogenesis for some years, certainly before *Daedalus*. In his first novel, *Crome Yellow*, published in 1921, one of his characters evokes a future where 'an impersonal generation will take the place of Nature's hideous system . . . In vast state incubators rows upon rows of gravid bottles will supply the world with the population it requires.' Quoted in Thody, 1973, p. 50.

73 Huxley, A., 1950, p. 94.

74 Ibid., p. 176.

75 Helen Parker stresses that Huxley's novel is the most pessimistic of six science fiction novels she discusses which envisage biological manipulation. It is also the earliest, and the best known of her examples. The others include Frank Herbert's *The Eyes of Heisenberg* (1966), Theodore Sturgeon's *Venus Plus X* (1966), Richard Cowper's *Clone* (1972) and Kate Wilhelm's *Where Late the Sweet Birds Sang* (1976). See Parker, 1984, Chapter 5.

76 On Huxley's relations with film as a new medium see Squier, 1994.

77 Gabor, 1964, quoted in Watt, 1975, p. 16.

78 For an account of the book's reception, and a selection of reprinted reviews, see the compilation edited by Watt (1975). He records that *Brave New World* was immediately Huxley's bestselling novel, selling 23,000 hardback copies in England in 1932–33 and 15,000 copies in the USA. The book still sells steadily.

79 West, 1932, reprinted ibid. No doubt West, H.G. Wells's paramour, knew of Haldane's assurance that none of the predictions in *Daedalus* were any more outlandish than Wells's, not Leonardo's, vision of the aeroplane.

80 Needham, 1932, reprinted ibid.; original emphasis. Not surprisingly, another favourable review came from Bertrand Russell. Russell's *The Scientific Outlook*, in which he developed the position first

put forward in *Icarus*, was a prime source for *Brave New World*.

81 The article was titled 'No father to guide them'. Ratcliff, 1937. For further details, see Bullough, 1994, pp. 133–4.

82 *Microbe Hunters*, first published in 1926 soon after De Kruif left the Rockefeller Institute, was a remarkable popular success. It passed through no fewer than 69 printings in four US editions between 1926 and 1945. De Kruif, 1940.

83 From the cover of Calder, 1961.

84 Reported in Crowther, 1970, p. 112.

85 Laurence, 1935.

86 Anon., 'Glass heart', *Time*, 26 (1 July 1935), pp. 41–2.

87 Corner, 1964, p. 236. The wide publicity for Carrel's work helped boost his now largely forgotten book, *Man the Unknown* later that year. An idiosyncratic blend of mysticism and apologetics for Fascism, this bizarre and fascinating book contributed one phrase to the debate I am tracing here which outlived Carrel's text. In his last chapter, Carrel speculates about the power of biology to transform the human race, declaring that 'to progress again man must remake himself. And he cannot remake himself without suffering. For he is both the marble and the sculptor.' Carrel, 1948, p. 252.

88 Kohler, 1976. As Lily Kay has recently emphasised, the Rockefeller programme was strikingly similar to the vision of Haldane and the Huxleys, envisioning a combined development of genetics, physiology, psychobiology, endocrinology and psychology into 'a new science of man', dedicated to rational control and improvement of human life. See Kay, 1993, especially pp. 43–6.

89 Jacques Monod, quoted in Judson, 1979, p. 212.

90 Anon., 'Life in the making', *New York Times*, 29 June 1935, p. 14. Compare with one of the first magazine articles

to mention Stanley's and other, similar experiments of the time, by G.W. Gray in *Harper's Magazine*. Entitled, 'Where life begins: search among the gene, the virus and the enzyme', it did not foreshadow any startling developments (Gray, 1937).

6   Time-bombing Our Descendants

1   For early reviews of these post-war changes, see Greenberg, 1967; Rose and Rose, 1970; Glass, 1960. On the shift in the centre of scientific activity to the US, see Ben-David, 1971.
2   Greenberg, 1967. The remarkable growth of the NIH continued into the 1980s. The NIH budget for the fiscal year 1980 was approved by Congress at $3.4 billion. This compared with the National Science Foundation's $966 million.
3   See Boyer, 1985. The definitive account of attitudes to nuclear science is Weart, 1988, which I discuss again below.
4   Weart, 1987.
5   Weart, 1989, pp. 5–6.
6   Ibid., p. 15.
7   Quoted ibid., p. 18.
8   Ibid., p. 67. See also Haynes, 1995, p. 197.
9   Quoted in Brians, 1984.
10 See, for example, the slim British volume, *Atomic Radiation Dangers – and What They Mean to You* (Heckstall-Smith, 1958).
11 For a sample, see Silverberg, 1977. In his introduction to one of the stories in this anthology, Anderson and Waldrop's notably bleak 'Tomorrow's Children', Silverberg remarks that 'For years after Hiroshima and Nagasaki, the pages of science fiction magazines were filled with stories describing human mutations that would come into the world after the atomic devas-

tation of World War III.' See also an earlier collection: Conklin, 1965, first published in 1955.
12 For a review of the whole history of the genre, see Hagar, 1982.
13 Carlson, 1967.
14 The best source on Müller is Carlson's biography (1981). See also Pontecorvo, 1968; Carlson, 1967, 1971; Sonneborn, 1968; Allen, 1970; and Lederberg's introduction to Müller's collected papers, Müller, 1962, which also contains a complete bibliography of his writings to that date.
15 Müller, unpublished autobiographical notes, quoted in Sonneborn, 1968. On Morgan's *Drosophila* research programme, and Müller's contribution, see Allen, 1969, 1978; Carlson, 1974, 1981; and Roll-Hansen, 1978.
16 In an unpublished paper read to a student society; Müller, 'Revelations of biology and their significance', cited by Sonneborn, 1968.
17 For a history, and a guide to the literature, see Kevles, 1986. For a broad reading of the history of twentieth-century medical genetics, see Turney and Balmer, in press.
18 Müller, 1927.
19 Carlson, 1981, p. 150.
20 Anon., 'Men and mice at Edinburgh: reports from the Genetics Congress', *Journal of Heredity*, 30 (1939), p. 371.
21 Müller, 1964b.
22 Müller, 1948, 1956.
23 Carlson, 1967.
24 Müller, 1926.
25 Judson, 1979, pp. 48–9.
26 Or, perhaps more to the point, a number of physicists became geneticists, of a kind.
27 Sonneborn, 1968.
28 See Weart, 1988, p. 4.
29 Crowther, 1970, p. 314.
30 Rosenberg, 1977. Thanks to Jamie Fleck for this reference.
31 Anon., 'The thinking machine', *Time*,

22 January 1949, p. 66. Anon., 'Paradise Lost', ibid., 6 August 1951, pp. 70–2. See also Walter, 1950.

32 Berkeley, 1949. See also Rosenberg, 1976, 1977 which extended this study into later material. Professor Rosenberg did not pursue this work on early reactions to artificial intelligence, 'mainly because of a fairly strong antipathy within the artificial intelligence community to this kind of analysis' (Rosenberg, personal communication). Plainly, a study closely related to mine could still be written about this history. See also Mazlish, 1993.

33 See, for example, the unremittingly technophobic *The Robots are Among Us* (Strehl, 1955).

34 Allen, 1975, p. 78. Note also that in 1956 Walter himself published a sub-Wellsian utopian novel in which, among other things, he predicted a society which had mastered human genetics and ectogenesis.

35 See Anon., 'Semi-creation', *Time*, 25 May 1953, p. 82; Fraenkel-Conrat, 1956.

36 Cited by Aldiss, 1975, p. 338. Another notable story of this period was Robert Heinlein's 'Jerry was a Man', in which a court must pronounce on whether an intelligent, genetically altered chimpanzee has rights. See Heinlein, 1953, cited by Sargent, 1976, p. 337. But this story was more remarkable for its embodiment of ideas more commonly found in the science fiction of the 1960s than for making any great impact when first published.

37 Note, though, the celebrated novel of social criticism by Pohl and Kornbluth, *The Space Merchants* (*Gravy Planet* in the US), which incorporates a sophisticated biological technology for food production based on a vast tissue-culture 'chicken little', a just recognisable descendant of Carrel's immortal chick-heart culture. Pohl and Kornbluth, 1952.

38 Lewontin, 1968. These warm words come from an anti-reductionist and latter-day critic of the Human Genome Project. See Lewontin, 1995.

7 Priming the Biological Time Bomb

1 Most fancifully in a major supplement on cell biology in the *National Geographic* in 1976, which included an account of how 'In April 1953 a 25 year-old American postgraduate student named James Dewey Watson and a British physicist, Francis Crick, dropped a biological bombshell, one that left newspaper editors around the world struggling to fit the words "deoxyribonucleic acid" into a catchy headline. Most opted to use the abbreviation DNA'. Gore and Dale, 1976, p. 368.

2 Calder, 1961; Yoxen, 1977, p. 347.

3 For overviews of the early history of molecular biology, see Judson, 1979; Yoxen, 1977; Allen, 1975, Chapter 7; Olby, 1974a; Portugal and Cohen, 1977. Later interpretations include Keller, 1992, and Kay, 1993.

4 Judson, 1979, p. 613.

5 Keller, 1992.

6 It is instructive to compare the breathless chronicling of the twists and turns of chaos and complexity theory in the 1990s, to see how the dynamics of science reporting have changed. Writers (and scientists) now go looking for the latest scientific revolution, and report it almost before it has taken shape.

7 Beck, 1961, p. 218.

8 Ibid. The passage recalls the influence of Schrödinger's *What is Life*, with its emphasis on the properties a gene must have, as an inspiration for a number of those who came into biology after World War II.

9 Ibid., p. 244.

10 Bronsted, 1956.

11 Rostand, 1959, p. 28. On Delage's work, see Chapter 4.

12 Rostand, 1955.

13 Rostand, 1959, p. 14; original emphasis.

14 Ibid., p. 21.

15 Ibid., p. 33.

16 Ibid., p. 68.

17 Ibid., p. 69.

18 Ibid., p. 91.

19 Pfeiffer, 1959.

20 Anon., 'The man of the future', *TLS*, 20 March 1959, p. 165.

21 Anon., *New Yorker*, 6 June 1959, p. 158.

22 Barnett, 1959.

23 Judson, 1979, p. 613.

24 Yoxen, 1977, p. 273.

25 These early stages in popularisation of molecular biology are described in detail in Yoxen, 1977, on which I draw heavily here. See especially Chapters 5 and 6.

26 Crick, 1957, quoted by Judson, 1979, p. 353.

27 Anon., *New York Times*, 2 February 1962, p. 1.

28 Margerison, 1962. See Yoxen, 1977, Chapter 5.

29 Galton, 1962.

30 Müller (1964a). Müller was a consistent advocate of the idea that the synthesis of DNA was to be equated with bio-genesis. See Müller, 1961a.

31 Anon., 'Probing heredity's secrets', *New York Times*, 12 September 1963, p. 36.

32 Ibid.

33 Seitz quoted in Anon., 'Hereditary control by man is foreseen', *New York Times*, 20 October 1963, p. 44; Abelson, 1965, reported ibid., 17 July 1965, p. 6; Glass, quoted in Anon., 'Vistas of genetics. AD 2,000', ibid., 19 February 1967, p. 8; Nirenberg, 1967, reported ibid., 12 August 1967, p. 12. Abelson was commenting on the symposium volume *The Control of Human Heredity and Evolution* (Sonneborn, 1965).

34 For example, Reinhold, 1967, 1968.

35 See Anon., 'The creation of life,' edito-rial, *New York Times*, 18 September 1965, p. 4. Kornberg, see Schmeck, 1967.

36 Yoxen, 1977, p. 260. See also Greenberg, 1967.

37 Kendrew, 1966. The book was based on a series of televised BBC lectures in 1964. Although Kendrew's Nobel-winning work was on protein structure, the DNA helix was the main element in his story, as the title suggests.

38 Lessing, 1967.

39 Ibid., p. 25.

40 Ibid., p. 51.

41 Leach, 1968.

42 Herbert, 1966.

43 Winstanley, 1976.

44 As Robert Olby argues, in an article nevertheless entitled 'The molecular revolution in biology'. Olby, 1990.

45 For the full story, see Bud, 1993.

46 Fleming, 1969; emphasis added.

47 The richness of Fleming's reading of the biological revolution may be contrasted with an essay from 1963, on the 'Revolution in biology', by Bentley Glass. He simply means the advance of knowledge, growing exponentially, and the educational problems this implies. There is no special focus on molecular biology, though Nirenberg's work on the genetic code gets a mention, and the piece covers the whole range of life sciences from evolution to ecology. The overall idea of a revolution is much more diffuse than the biological revolution seen by Fleming and other writers later in the 1960s. So we can say that the more developed idea appeared some time in the five years between 1963 and 1968. Glass, 1963.

48 Anon., 'Control of life', *Life*, 39, No. 6 (20 September 1965), pp. 53–72.

49 Anon., 'Gift of life from the dead', *Life*, 39, No. 7 (4 October 1965), pp. 58–72; Anon., 'Rebuilt people', *Life*, 39, No. 8 (18 October 1965), pp. 58–78.

50 Rosenfeld, 1965 (part four of the series cited above).

51 Fox and Swazey, 1974. The book is a study of heart and kidney transplantation and dialysis in the 1960s by a sociologist and a historian.

52 Ibid., p. 31.

53 Ibid., pp. 149–211.

54 Taylor's next book, a jeremiad if ever there was one, was *The Doomsday Book*, on environmental questions.

55 Waddington, 1968.

56 See, for example, Rorvik, 1978a; Packard, 1978.

57 Taylor, 1968, pp. 11–12. All quotations are taken from the Panther paperback edition of 1969.

58 Ibid., p. 76.

59 Ibid., p. 103; original emphasis.

60 Ibid., p. 76.

61 See especially Wolstenholme, 1963, and Sonneborn, 1965. The CIBA Foundation volume edited by Wolstenholme has papers by Lederberg and Haldane – one of his last publications – among others which were quoted many times by later authors. Fleming's article, for example, makes extensive use of selective quotation from this volume.

62 The quote from Diderot concerns the hero viewing 'a warm room with the floor covered with little pots and on each pot a label; soldiers, magistrates, philosophers, poets, potted courtesans, potted kings'. Taylor, 1968, p. 11; Rostand, 1959, p. 15. The quote from Bacon, whom only Rostand names, is, 'knowledge without the corrective of charity, has some nature of venom or malignity'. Taylor, 1968, p. 245, Rostand, p. 70.

63 See, for example, Taylor, 1968, pp. 33, 60, 135, 168, 197, and, correspondingly, Rostand, 1959, pp. 11, 34, 67, 88, 65–6.

64 The one point on which Taylor takes issue with the optimistic predictions of future success is in discussion of the creation of life, which he feels is very far off. But he regards this as less important than genetic engineering, which he believes will come much sooner.

65 Taylor, 1968, p. 188.

66 Ibid., p. 245.

67 Anon., 'Today's trees tomorrow', *TLS*, 25 April 1968, p. 411.

68 Storr, 1968; Morison, 1968; Steiner, 1968.

69 See Anon., 'Green fruits of knowledge', *Economist*, 25 April 1968, p. ix; Calder, R., 1968; Waddington, 1968, 1969; Calder, N., 1968.

70 Calder, R. 1968; Anon., *Economist* (op. cit.).

8   The Baby of the Century

1 Wood, 1969, p. 161. On the history of reproductive physiology see also Parkes, 1966.

2 The historical details which follow are taken from a number of secondary accounts: Vaughan, 1970; Wood, 1969; Grossman, 1971; Packard, 1978; Edwards and Steptoe, 1980. My interest here is in the broad outline of the subject, so I have turned to primary sources only to resolve a few minor contradictions between these accounts.

3 Anon., *Science Newsletter*, 27 October 1951, cited by Williams, 1978, p. 79.

4 Taylor, 1968, p. 26.

5 Anon., 'Scientists grow a human embryo', *New York Times*, 14 January 1961, p. 21.

6 See Rorvik, 1978a, pp. 80–1; Francoeur, 1973, pp. 57–9; Packard, 1978, pp. 192, 213–14; Grossman, 1971. Packard gives the date of the announcement as 1959.

7 Rorvik, 1978a, p. 80.

8 Grossman, 1971; Packard, 1978, p. 205.

9 Müller, 1961a.

10 Balio, 1967; Atkinson, 1961. The latter

is a sceptical article from the *New York Times* entitled 'Some drawbacks are found in Professor Müller's plan for breeding people'.

11 Smith, W., 1964.

12 Taylor, 1968, p. 34.

13 Packard, 1978, p. 201; Rorvik, 1978a, p. 32.

14 Tucker, 1966.

15 Edwards, 1969.

16 Anon., 'What comes after fertilisation?', *Nature*, 221 (15 February 1969), p. 613.

17 *The Times*, Friday, 14 February 1969, p. 1.

18 Smith, A., 1969.

19 Anon., 'New hope for the childless', *Guardian*, 14 February 1969, p. 1.

20 Tucker, 1969.

21 Compare the more light-hearted cartoon in the *Morning Star*, in which a woman reading a newspaper whose headline proclaims the advent of the test-tube baby leans across the breakfast table to her husband with the words 'I hate to tell you Fred, but you've just been made redundant'. *Morning Star*, 15 February 1969, p. 2.

22 Anon., 'Test-tube fertility hope for women', *The Times*, 15 February 1969, p. 1.

23 Anon., 'Implications of the test-tube embryo', ibid., p. 8.

24 Jessel, 1969.

25 Anon., 'Life in the test-tube', *The Times*, 15 February 1969, p. 9.

26 Brunner, 1970, quoted in Scholes and Rabkin, 1977, p. 81.

27 Davy, J. (1966), reproduced in Brunner, 1969, p. 247.

28 Macmillan, 1969.

29 Gurdon's classic experiments were important for developmental genetics because they were regarded as the definitive demonstration that all body cells contain the full complement of genes. They were the pretext for much discussion of cloning, a symbol of bio-logical manipulation almost as potent as the test-tube baby.

30 Anon., 'How sinister is biology?', *Nature*, 1 March 1970, p. 803.

31 Maddox, 1969.

32 Bedford, 1970.

33 Wood, 1970.

34 Chartres, 1970.

35 Anon., 'Mr Steptoe's babies' (2nd leader), *The Times*, 27 February 1970, p. 9.

36 Interview with Kit Pedler, *Radio Times*, 5 February 1970, p. 2.

37 Remarks from a BBC Radio interview, subsequently quoted in *The Times*, in Wright, 1970. Pedler, who was head of the anatomy department at London University's Institute of Ophthalmology, was also a prominent antivivisectionist and, latterly, anti-nuclear campaigner.

38 Cardinal Gordon Gray, quoted in the *Daily Mirror*, 25 February 1970.

39 Bateman, 1969.

40 Anon., 'The moral problem' (editorial), *Daily Express*, 25 February 1970.

41 Pincher, 1970, and Cummings cartoon, *Daily Express*, 25 February 1970, p. 8.

42 Anon., 'Incubator mothers forecast', *The Times*, 26 February 1970, p. 1.

43 Leach, 1970.

44 A viewpoint developed at greater length, and with some cogency, in Leach's book *The Biocrats* (1972).

45 Silcock, 1970.

46 Le Palourel, 1970.

47 Murray, 1970.

48 Leach, 1970.

49 A less subtle, though somewhat cryptic, link between test-tube babies and heart transplants was made by the reporter who managed to extract a quote which justified the following comment in *The Times*: 'Dr Christian Barnard said in London yesterday that he did not think "test-tube" baby research would lead to the creation of a Frankenstein Monster'. Readers at the time would

have known that Barnard performed the first human heart transplant. Even so, this was taking the power of a phrase to evoke a script to an unusual extreme. Comment appended to Anon., 'Incubator mothers forecast', *The Times*, 26 February 1970, p. 1.

50 The growth of the field was underlined by publication of a four-volume *Encyclopedia of Bioethics*, produced under the auspices of the Kennedy Center for Bioethics, Georgetown University, in 1979 (Reich, 1979).

51 Kass, 1971a; original emphasis.

52 Lappe, 1972, cited by Reilly, 1987, p. 209.

53 Ramsey, 1972a; original emphasis.

54 For example, Fletcher, 1974.

55 Kass, 1971b.

56 Ibid.; original emphasis.

57 Ramsey, 1972a.

58 Ramsey, 1972b.

59 Anon., 'Genetic engineering: reprise' (editorial), *JAMA*, 220 (5 June 1972), pp. 1356–7.

60 See, for example, Anon., 'Man into Superman. The promise and peril of the new genetics', *Time*, 19 April 1971; Anon., 'AMA urges delay on test-tube baby', *New York Times*, 1 May 1972, p. 39; Anon., 'Storm over work on test tube babies' *The Times*, 18 October, 1971, p. 5.

61 Anon., 'Storm over work', op. cit.

62 Anon., 'Deformity fears over test tube babies', *The Times*, 20 October 1971. The thalidomide disaster, which the US escaped, was a leading example of technology biting back through the 1960s and 1970s in Britain. Nevertheless, it was in Britain, not the US, that *in vitro* fertilisation work was allowed to go ahead at this time.

63 Small wonder that when reporting a later piece of research on cloning, *Time* magazine suggested that 'ethicists and science-fiction writers have almost indistinguishable job descriptions'.

Lemonick, 1993. See my discussion of fact and fiction in Chapter 10.

64 Gaylin, 1972.

65 Examples most widely cited at the time were the Tuskegee syphilis study, and injection of live cancer cells into elderly chronic patients at a Brooklyn hospital. See, for example, Packard, 1978, p. 333.

66 Marx, 1973.

67 Chedd, 1974.

68 Quoted in Reilly 1987.

69 For example, a claim by Professor Bevis of Leeds University in July 1974. Speaking at the annual meeting of the British Medical Association, he declared that there was at least one test-tube baby alive in Britain, and two elsewhere in Europe. He declared himself surprised at the ensuing publicity, despite having told a CIBA Foundation symposium the previous year that 'the mass media are hysterically interested in embryo transfer'. When pressed, he announced he was withdrawing from the field, and that he had refrained from publication to protect the children. He was not heard from on the subject again. See Anon., 'Test-tube baby alive and well in Britain', *The Times*, 16 July 1974; Packard, 1978, pp. 189–90; Wolstenholme, 1973, p. 95.

70 Austin and O'Reilly, 1978; Jennings, 1978.

71 See Nally, 1978, which focuses on the arrival in Oldham of a five-man team from the US *National Inquirer*, and their free-spending efforts to discover the parents' identity and whereabouts.

72 Both the parents and Steptoe and Edwards played down the sums involved when their separate accounts of the birth were published. Brown and Brown, 1979; Edwards and Steptoe, 1980.

73 As a headline in the *Manchester Evening News* put it. Harris, 1978.

74 For example, Hodgkinson, 1978.

75 Eggington, 1978.

76 Anon., 'All about that baby', *Newsweek*, 7 August 1978.

77 Anon., 'In vitro infant raises tempest in test tube', *Science*, 201 (11 August 1978), p. 510.

78 Anon., 'Reproductive technology: whose baby?', *Nature*, 3 August 1978, p. 409.

79 Anon., 'The joy, the hope and the disquiet', *Daily Mail*, 27 July 1978.

80 Gale, 1978.

81 Tucker, 1978.

82 Anon., 'Bypassing a block to conception', *The Times*, 27 July 1978, p. 15.

83 Brown, 1978.

84 Quoted in Anon., 'To fool (or not) with Mother Nature', *Time*, 31 July 1978, p. 52.

85 Gwynne, 1978, p. 44.

86 Golden, 1978, p. 46.

87 Ibid.

88 Gwynne, 1978.

89 Longo, 1976; Charbonneur, 1976. One critic, at least, would not have been unhappy to see the first *in vitro* fertilisation produce a defective child. Paul Ramsey had written in 1972 in the *Bulletin of the Atomic Scientists*, 'Perhaps one can express the paradoxical and macabre "hope" that the first example of the production of a child by genetic engineering of its parents' gametes will prove to be a bad result – and that it will be well advertised, not hidden from view'. Ramsey, 1972, reprinted in Mertens, 1975.

90 More than a year later, the Browns told the *Manchester Evening News* how it was amazing that 'many people expect her [Louise] to be in some way different'. The same article declared how 15-month-old Louise appeared 'miraculously normal' during a walk around the centre of the city. Bullas, 1979.

91 Francouer, 1972, pp. 112–16. A smaller poll in the vicinity of Stanford University in 1973 gave very similar results (Miller, 1974).

9   The Gene Wars: Regulating Recombinant DNA

1 The most important accounts are Krimsky (1982) for the US, Bennett et al. (1986) for the UK, and Wright (1994), which offers a sophisticated comparison of the course of the debate in several countries.

2 Berg et al., 1974.

3 Singer and Soll, 1973.

4 Wade, 1974.

5 See Krimsky, 1982, pp. 30–3.

6 The most compelling account of the meeting remains Michael Rogers' report for *Rolling Stone* magazine, also included in his book on the recombinant DNA debate, *Biohazard*. Rogers, 1975, 1977.

7 Cohen, 1977.

8 Jukes, 1976.

9 Quoted in Anon., *The Canadian*, 2 July 1977.

10 May, 1977.

11 Davis, 1976; Watson, 1977.

12 Fraenkel-Conrat, 1975. Compare Steiner, 1978, a longer and more erudite treatment of the same conflict, which I discuss in the next chapter.

13 For example, Goodfield, 1977; Nelkin, 1978.

14 See Nelkin, 1995 for a longer discussion of resistance to new technology which contrasts biotechnology and information technology.

15 Gaylin, 1972, as discussed in the previous chapter. See Krimsky, 1982, p. 165 for more about Gaylin's background. Cloning was also becoming a staple of biological science fiction in the 1970s: see especially Sargent, 1976, a novel which describes the lifting of a moratorium on biological research, thus permitting experimental human cloning. The facts and fiction of cloning are discussed in more detail in the following chapter.

16 Anon., *US News and World Report*, 82 (11 April 1977); Anon., *Time*, 109 (18 April 1977); Anon., *Newsweek*, 16 August 1976.

17 Rogers, 1975, p. 208.

18 Chargaff, 1976.

19 Sinsheimer, 1975. Sinsheimer's particular concern at that time was the possible existence of a 'species barrier' to DNA transfer in nature, which recombinant DNA might breach. In the late 1980s, as chancellor of the University of California at Santa Cruz, he was an early and forceful advocate of the Human Genome Project.

20 Lubow, 1977, reprinted in Watson and Tooze, 1981, pp. 118–26.

21 Lewin, 1977a.

22 Quoted in Rogers, 1977, p. 210. See also Mendelsohn, 1979, 1989.

23 Krimsky, 1982, p. 307. Before producing the most detailed early historical account of the controversy in the US, Krimsky was a member of the Cambridge Review Board.

24 Quoted ibid.

25 The main title of the essay is 'Frankenstein at Harvard'. Mendelsohn, 1979.

26 Ibid., p. 327.

27 Bennett et al., 1986, p. 10.

28 Ashby, 1975, p. 3.

29 Bennett et al., 1986, p. 160.

30 Ibid.

10 Conclusion: The Human Body Shop

1 Notable examples include the British ATV company's film, starring Robert Powell, Carrie Fisher and Sir John Gielgud, aired in 1984, and still sold on video in the mid-1990s.

2 See Schelde, 1993 for many examples.

3 Branagh, 1994a.

4 Ibid., p. 19. Branagh's depiction of Victor Frankenstein is thus a major counter-example to Christopher Toumey's contention that 'mad scientists' have consistently been rendered more amoral as nineteenth-century texts have been adapted to make twentieth-century films, though in fairness his film does post-date Toumey's paper, if not the later book in which he repeats the argument. See Toumey, 1992 and 1996.

5 For example, the literary scholar Roslynn Haynes comments on Branagh's film that 'it pays scant attention to the technical and sociological constraints of the alleged time and setting of the novel unless they can be exploited in the cause of violence. Scenes such as the exodus from a plague-ridden Ingolstadt ... not only overshadow the intellectual and moral issues of the novel, but actively subvert many of its finer points.' Haynes, 1995.

6 Spin-offs included a lavishly illustrated version of the screenplay (Branagh, 1994), and, separately, a novelisation of the film. The latter, a notably plodding book by Leonore Fleischer, is confusingly entitled *Mary Shelley's Frankenstein* (Fleischer, 1994). Branagh also recorded an audiobook version of Mary Shelley's text.

7 Also worth noting in passing is John Frankenheimer's new film of *The Island of Doctor Moreau* (1996). This is already famous more for its cinematic ineptitude than for attracting viewers, but it falls neatly in line with Brian Aldiss's suggestion that a modern-day Moreau would be a genetic engineer. The beast-people of this latter-day Moreau, who was played by Marlon Brando, are 'animals that have been fused with human genes'. A similar fusion besets Brundle in David Cronenberg's remake of *The Fly* (1986), in which the monster at the centre of the action is created from joining human and fly genes through 'integration at the molecular genetic level'.

8   Quoted in McGregor, 1993.

9   Jaynes, 1995. It is not clear to me what validity figures like this have; only that similar claims are made for other, later films. As this book was completed, the film of *The Lost World* was reportedly the fastest-grossing release in history in the USA.

10 Just as Crichton saw the appeal of recreating extinct creatures, so film-makers have repeatedly relished the technical challenge of bringing dinosaurs to the screen. The first notable dinosaur film was *The Lost World* (1925), based on the Conan Doyle story of the same name. This was remade in 1960, six years before *One Million Years BC*, itself a remake of a 1940 epic, and the film which led many people to believe that dinosaurs and early humans co-existed. There are also, of course, numerous monster films featuring dinosaurs, notably the Japanese Godzilla features. All of these films feature animated monsters. Spielberg combined animatronic creatures and computer-generated images.

11 On the quagga, see Hart, 1988, which discusses at length the making of a British television programme about the possibility of recovering fragments of quagga DNA in the mid-1980s.

12 For possible explanations of 'dino-mania' see Gould, 1993.

13 Ibid., 1993.

14 Quoted in a tabloid newspaper feature, liberally illustrated with colour stills from the film, which describes the story as 'closer to the truth than most people would imagine'. Burke, 1991.

15 Stableford, 1993, p. 9.

16 Anon., 'Reproductive technology: whose baby?', *Nature*, 274 (3 August, 1978), p. 409; emphasis added.

17 Quoted from an exchange with the sociologist Amitai Etzioni in the *New York Times* in September 1970, following Etzioni's suggestion that 1 per cent

of the $10 million Lederberg had sought for a 'National Genetics Task Force' should be set aside for funding of research on the social consequences of genetic engineering. Twenty years later, this is exactly what Congress agreed to for the budget for the Human Genome Project. As related in Etzioni, 1973, p. 50.

18 Abelson, 1965.

19 Abelson, 1971.

20 Nirenberg, 1967. For Singer's views see Greenberg, 1967, and Singer, 1977a.

21 Watson, 1971.

22 See quotes in Anon., 'A test tube baby is not a clone', *Time*, 31 July 1978, p. 50.

23 I refer to the meetings at Asilomar in 1975, Oxford in 1975 and Wye College in Kent in 1979. The attempt was only successful in the case of the last of these. See Lewin, 1979.

24 See Nelkin and Lindee, 1995.

25 Baltimore, 1978b.

26 See Chapter 8, p. 159.

27 Steiner, 1978.

28 There are exceptions, of course, most obviously *Dr Jekyll and Mr Hyde*.

29 Bromhall, 1978.

30 Rorvik, 1978a, p. 181.

31 Easton, 1978. This review, in a science fiction journal, concluded that Rorvik deserved 'far more praise than blame' for making bioethics 'accessible'.

32 Rorvik, D., 1978b. This book closely paralleled Taylor's earlier volume. Although classified as non-fiction it contained extended passages of speculation, presented as such on this occasion. Articles on Rorvik's later book include Cooper, 1978, Culliton, 1978a, 1978b.

33 The book was Ebon, 1978, the film *The Clone Affair*, broadcast on BBC1 on 15 May 1979.

34 Durden-Smith, 1979. Rorvik's book also stimulated additional publicity for the film of Ira Levin's *The Boys from Brazil*, released later that year. The

novel postulates a renegade Nazi geneticist (the notorious Dr Mengele) attempting to bring a number of Hitler clones to adulthood in the post-war world. Derek Bromhall, apparently unconcerned by explicitly fictional representations of his work, acted as technical adviser during the making of the film.

35 Commander Raft, in *Beloved Son* (Turner, 1978).

36 Elmer-Dewitt, 1993. The *Newsweek* report is Adler, 1993. Both were cover stories.

37 Wilmut et al., 1997.

38 Woodward, 1997.

39 Kluger, 1997.

40 Coupland, 1997.

41 In a public debate in London on 23 March 1995 under the banner 'Monster Myths – Are writers demonising the new genetics?'

42 McKernan, 1995.

43 Even if we assume a passive audience, which simply bears the impress of the last thing they saw, simple content analysis would not indicate that they would all believe that biologists are no more than Frankensteinian figures. This story is ever-present, as this book has indicated, but that does not mean that it is always, or even very often, the main image of real science which is available. See the discussion of the embryo research debate below for a clear example of this.

44 Nelkin, 1987. See also Einseidel, 1992.

45 McLeod et al., 1991; Moores, 1993.

46 Mulkay, 1996. I am grateful to Professor Mulkay for allowing me to see an early version of this paper before publication. Edwards's comments, quoted by Mulkay, are in his book reflecting on the embryo debate (Edwards, 1989).

47 As one MP put it: 'I cannot help feeling that Mary Shelley's spectre of Dr Frankenstein's monster impinges heavily on our subconscious when we address ourselves to the problem of embryology, causing a fear and revulsion against the possible products of the ruthless pursuit of knowledge for its own sake or the application of medical techniques to create monsters or superhumans'. Dobson, quoted in Mulkay, 1996.

48 Evelyn Fox Keller's phrase. See Keller, 1992.

49 Nelkin and Lindee, 1995, p. 198.

50 Gilbert, 1995, p. 571. The advent of the genetic body is discussed further in Turney and Balmer, in press.

51 Rheinberger, 1995.

52 Ladenis, 1996; Schoon, 1996.

53 The philosopher and biophysicist Bernard Rollin makes the same point in his recent book on genetic engineering of animals, *The Frankenstein Syndrome* (1995), in which he goes on to discuss in detail how to take that particular debate on to more tractable terrain.

54 We have to escape the syndrome correctly diagnosed by William Leiss: 'The inner desire and terror that characterize humanity's oldest experiences with technology continue to feed a kind of fatalism whereby people gratefully accept the fruits of human ingenuity while dreading the eruptions of uncontrollable malevolence from its handiwork'. Leiss, 1972, p. 29.

55 Brian Stableford makes a similar point: 'We all tell ourselves futuristic stories all the time; they are inherent in our daydreams and our career plans, and in our anxious fears for our own old age and adulthood of our children. When the pace of social and technological change accelerates, we all need more help and more practice in writing the scripts for these stories, which are part and parcel of the narratives of our real lives'. Stableford, 1993, p. 7.

Bibliography

This is an alphabetical listing of books and articles consulted.
Place of publication is London unless otherwise stated.

Abelson, P. (1965) 'Fruit from the tree of knowledge' (editorial), *Science*, 149, p. 251.

Abelson, P. (1971) 'Anxiety about genetic engineering' (editorial), *Science*, 171, p. 285.

Adams, C. (1990) *The Sexual Politics of Meat.* Cambridge, Polity Press.

Adler, J. (1993) 'Clone Hype', *Newsweek*, 8 November, p. 44.

Aldiss, B. (1973) *Frankenstein Unbound.* Jonathan Cape.

Aldiss, B. (1975) *Billion Year Spree.* Corgi Books.

Aldiss, B. (1990) *Introduction* to *The Island of Dr Moreau*, Everyman.

Allaby, M. (1995) *Facing the Future: The Case for Science.* Bloomsbury.

Allen, G. (1969) 'T.H. Morgan and the emergence of a new American biology', *Quarterly Reviews of Biology*, 44, p. 168.

Allen, G. (1970) 'Science and society in the eugenic thought of H.J. Müller', *Bioscience*, 20, p. 346.

Allen, G. (1975) *Life Science in the Twentieth Century.* Chichester, John Wiley.

Allen, G. (1978) *Thomas Hunt Morgan: The Man and His Science.* Princeton, NJ, Princeton University Press.

Anderson, R. (1978) 'One fine day', *New Scientist*, 77, p. 148.

Anobile, R. (ed.) (1974) *James Whale's 'Frankenstein'.* Pan Books.

Anon., 'All about that baby', *Newsweek*, 7 August 1978.

Anon., 'AMA urges delay on test-tube baby', *New York Times*, 1 May 1972, p. 39.

Anon., 'A test-tube baby is not a clone', *Time*, 31 July 1978, p. 50.

Anon., 'Biologists hopeful of solving secrets of heredity this year', *New York Times*, 2 February 1962, p. 1.

Anon., 'Biology's "atom bomb"', *Christian Science Monitor*, 26 January 1976.

Anon., 'Bypassing a block to conception', *The Times*, 27 July 1978, p. 15.

Anon., 'Clone hype' (cover story), *Newsweek*, 8 November 1993, pp. 44–9.

Anon., 'Close to mystery of life, says Loeb', *New York Times*, 21 July 1912.

Anon., 'Control of life', *Life*, 20 September 1965, pp. 53–72.

Anon., 'Creating new forms of life', *US News and World Report*, 11 April 1977, p. 80.

Anon., 'Creation by chemistry', *Daily Mirror*, 6 September 1912.

Anon., 'Deformity fears over test tube babies', *The Times*, 20 October 1971.

Anon., editorial, *The Nation*, 74, 16 January 1902, p. 46.

Anon., editorial, *New York Times*, 26 September 1907.

Anon., editorial, *Scientific American Monthly*, March 1921, p. 221.

Anon., 'Frontiers of science still subject to fear of the unknown', *Euroforum*, 11 May 1979.

Anon., 'Genesis by lightning', *Scientific American*, July 1953, p. 42.

Anon., 'Genetic engineering: reprise', *Journal of the American Medical Association*, 220, 5 June 1972, pp. 1356–7.

Anon., 'Gift of life from the dead', *Life*, 4 October 1965, pp. 58–72.

Anon., 'Glass heart', *Time*, 1 July 1935, pp. 41–2.

Anon., 'Green fruits of knowledge', *Economist*, 25 April 1968.

Anon., 'Head of Pasteur Institute proposes to replace electric chair by laboratory', *New York Sun*, 8 December 1912.

Anon., 'Hereditary control by man is foreseen', *New York Times*, 20 October 1963, p. 44.

Anon., 'How sinister is biology?' *Nature*, 225, 1970, p. 803.

Anon., 'Implications of the test-tube embryo', *The Times*, 15 February 1969, p. 8.

Anon., '*In vitro* infant raises tempest in test-tube', *Science*, 201, 11 August 1978, p. 510.

Anon., 'Incubator mothers forecast', *The Times*, 26 February 1970, p. 1.

Anon., 'Life and Current Science' (editorial), *Independent*, 54, 1902, pp. 2265–6.

Anon., 'Life as per recipe' (editorial), *Daily Herald*, 6 September 1912.

Anon., 'Life still hides its secret' (editorial), *New York Times*, 9 August 1911.

Anon., 'Life in the test-tube', *The Times*, 15 February 1969, p. 9.

Anon., 'Man into Superman. The promise and peril of the new genetics', *Time*, 19 April 1971.

Anon., 'Mice and Men at Edinburgh: Reports from the Genetics Congress', *Journal of Heredity*, 1939, 30, p. 371.

Anon., 'Mr Steptoe's babies', *The Times*, 27 February 1970, p. 9.

Anon., 'New hope for the childless', *The Guardian*, 14 February 1969, p. 1.

Anon., 'New step toward test-tube babies', *The Times*, 14 February 1969, p. 1.

Anon., 'Now every woman can have a baby', *Manchester Evening News*, 14 February 1969, p. 1.

Anon., 'Origin of life' (leader), *Guardian*, 5 September 1912.

Anon., 'Paradise lost', *Time*, 6 August 1951, pp. 70, 72.

Anon., 'People plants', *Newsweek*, 16 August 1976.

Anon., 'Premature birth of test tube baby', *Nature*, 225, 1970, p. 886.

Anon., 'Probing heredity's secrets', *New York Times*, 12 September 1963, p. 36.

Anon., 'Professor Hill's task', *Daily Mirror*, 15 September 1928.

Anon., 'Prophecy that scientists will create life', *Daily Mirror*, 12 September 1928, p. 3.

Anon., 'Public faith in science stays high', *Nature*, 381, 1996, p. 355.

Anon., 'Reality and physical science', *The Times*, 13 September 1912.

Anon., 'Rebuilt people', *Life*, 18 October 1965, pp. 58–78.

Anon., *Report of the Working Party on the Experimental Manipulation of the Genetic Composition of Micro-Organisms* (Ashby Report). HMSO, Cmnd 5880, 1975.

Anon., 'Reproductive technology: whose baby?', *Nature*, 3 August 1978, p. 409.

Anon., review of *Rostand* (1959), *New Yorker*, 6 June 1959, p. 158.

Anon., 'Scientist protests', *Daily Mirror*, 14 September 1928.

Anon., 'Scientists and the secret of life', *Daily Mirror*, 5 September 1912.

Anon., 'Scientists grow a human embryo', *New York Times*, 14 January 1961, p. 21.

Anon., 'Sees way to control growth of tissues', *New York Times*, 21 May 1911.

Anon., 'Semi-creation', *Time*, 25 May 1953, p. 82.

Anon., 'Steptoe's shifting press relations', *New Scientist*, 9 November 1978.

Anon., 'Storm over work on test tube babies', *The Times*, 18 October 1971, p. 5.

Anon., 'Synthetic rubberism' (leader), *Daily Mail*, 6 September 1912.

Anon., '"Test-tube" baby alive and well in Britain', *The Times*, 16 July 1974, p. 1.

Anon., 'Test-tube fertility hope for women', *The Times*, 15 February 1969, p. 1.

Anon., 'The joy, the hope and the disquiet', *Daily Mail*, 27 July 1978.

Anon., 'The man of the future' (review of Rostand, 1959), *Times Literary Supplement*, 20 March 1959, p. 165.

Anon., 'The moral problem' (editorial), *Daily Express*, 25 February 1970, p. 8.

Anon., 'The thinking machine', *Time*, 22 January 1949, p. 66.

Anon., 'Tinkering with life', *Time*, 18 April 1977, pp. 32–45.

Anon., 'Today's trees tomorrow' (review of Taylor, 1968), *Times Literary Supplement*, 25 April 1968.

Anon., 'To fool (or not) with Mother Nature', *Time*, 31 July 1978, p. 52.

Anon., 'Vistas of genetics, AD 2000', *New York Times*, 19 February 1967, p. 8.

Anon., 'What comes after fertilisation?', *Nature*, 221, 15 February 1969, p. 613.

Atkinson, B. (1961) 'Critic at large – some drawbacks are found in Professor Müller's plan for breeding people', *New York Times*, 18 July 1961, p. 20.

Augenstein, L. (1969) *Come, Let Us Play God*. New York, Harper & Row.

Austin, F. and O'Reilly, P. (1978) 'First test-tube baby in July!', *Daily Mirror*, 21 April, p. 1.

Austoker, J. and Bryder, L. (1989) *Historical Perspectives on the Role of the MRC*. Oxford University Press.

Ayala, F. and Dobzhansky, T. (1974) *Studies in the Philosophy of Biology*. Macmillan.

Baldick, C. (1987) *In Frankenstein's Shadow: Myth, Monstrosity and Nineteenth Century Writing*. Oxford, Clarendon Press.

Balio, T. (1967) 'The public confrontation of H.J. Müller', *Bulletin of the Atomic Scientists*, 23, pp. 8–12.

Baltimore, D. (1978a) 'Genetic scaremongering', *Nature*, 272, p. 766.

Baltimore, D. (1978b) 'Limiting science: a biologist's perspective', *Daedalus*, Spring, pp. 37–45.

Bann, S. (ed.) (1995) *Frankenstein, Creation and Monstrosity*. Reaktion Books.

Barnes, B. (ed.) (1972) *Sociology of Science*. Harmondsworth, Penguin.

Barnett, A. (1959) 'The proper study', *New Statesman*, 14 March 1959, p. 378.

Barns, I. (1990) 'Cinematic resolutions of the "Frankenstein Problem"', *Science as Culture*, 9, pp. 7–48.

Basalla, G. (1976) 'Pop science: the depiction of science in popular culture', pp. 261–78 in G. Holton and W. Blanpied (eds) *Science and its Public: The Changing Relationship* (Boston Studies in the Philosophy of Science, Vol. 33). Dordrecht, Reidel.

Bastian, H. (1878) 'Spontaneous generation – a reply', *Nineteenth Century*, 3, p. 261.

Bateman, M. (1969) 'East, West, which manipulators work best?', *Sunday Times*, 1 March.

Bauer, M. (1994) 'Cultural segmentation and the resistance function of social representation: observations on public understanding of science'. Paper given at British Psychological Society, Social Psychology Section annual conference, September.

Bauer, M. (ed.) (1995) *Resistance to New Technology: Nuclear Power, Information Technology and Biotechnology*. Cambridge University Press.

Beadle, G. and Beadle, M. (1966) *The Language of Life*. Gollancz.

Beck, W. (1961) *Modern Science and the Nature of Life*. Harmondsworth, Penguin.

Beckwith, J. (1974) 'Social and political uses of genetics in the USA – past and present', in Lappe and Morison, 1974.

Bedford, R. (1970) 'Wife is waiting for test-tube baby', *Daily Mirror*, 24 February, p. 1.

Ben-David, J. (1971) *The Scientist's Role in Society*. Englewood Cliffs, NJ, Prentice-Hall.

Bennett, D., Glasner, P. and Travis, D. (1986) *The Politics of Uncertainty: Regulating Recombinant DNA Research in Britain*. Routledge.

Benthall, J. (1976) *The Body Electric: Patterns of Western Industrial Culture*. Thames & Hudson.

Berg, P. et al. (1974) 'Potential biohazards of recombinant DNA molecules', *Science*, 185, p. 303. Also in *Nature*, 250, p. 175, 1974.

Bergonzi, B. (1961) *The Early H.G. Wells*. Manchester University Press.

Berkeley, E. (1949) *Giant Brains, or Machines That Think*. Chichester, John Wiley.

Berman, M. (1983) *All That Is Solid Melts Into Air: The Experience of Modernity*. Verso.

Bernard, C. (1957) *An Introduction to the Study of Experimental Medicine*. New York, Dover.

Bigsby, C. (ed.) (1976) *Approaches to Popular Culture*. Edward Arnold.

Birkenhead, Lord. (F.E. Smith) (1930) *Life in the Year 2000*. Hodder & Stoughton.

Blank, R. and Bonnicksen, A. (eds) (1994) *Medicine Unbound: The Human Body and the Limits of Medical Intervention*. New York, Columbia University Press.

Bloom, H. (1987) *Mary Shelley's 'Frankenstein': Modern Critical Interpretations*. New York, Chelsea House.

Bloomfield, B. and Vurdabakis, T. (1995) 'Disrupted boundaries: new reproductive technologies and the language of anxiety and expectation', *Social Studies of Science*, 25, pp. 533–51.

Boyer, P. (1985) 'Culture and consciousness in the early atomic era', Chapter 7 in *By the Bomb's Early Light: American Thought and Culture at the Dawn of the Atomic Age*. New York, Pantheon Books.

X Bozzetto, R. (1993) 'Moreau's tragi-farcical island', *Science Fiction Studies*, 20, pp. 35–44.

Branagh, K. (1994a) 'Director's note – a tale for all time', in *Mary Shelley's 'Frankenstein'*. Pan Books.

Branagh, K. (1994b) 'Introduction – Frankenstein reimagined', in *Mary Shelley's 'Frankenstein'*. Pan Books.

Brians, P. (1984) 'Nuclear war in science fiction, 1945–59', *Science Fiction Studies*, 11, pp. 253–63.

Brierley, J. (1964) 'The biological sciences curriculum study publications', *Science*, 163, pp. 668–70.

Broks, P. (1993) 'Science, media and culture: British magazines, 1890–1914', *Public Understanding of Science*, 2, pp. 123–40.

Bromhall, D. (1978) 'The great cloning hoax', *New Statesman*, 2 June, p. 734.

Bronsted, H. (1956) 'The warning and promise of experimental embryology', *Bulletin of the Atomic Scientists*, March.

Brosnan, J. (1991) *The Primal Screen: A History of Science Fiction Film*. Orbit.

Brown, D. (1978) 'Genetic experiment fears', *Guardian*, 27 July.

Brown, L. and Brown, J. (1979) *Our Miracle Called Louise*. Paddington Press.

Brunner, J. (1969) *Stand on Zanzibar*. New York, Ballantine paperback.

Brunner, J. (1970) 'The genesis of *Stand on Zanzibar*', *Extrapolation*, 11/12.

Bud, R. (1993) *The Uses of Life: A History of Biotechnology*. Cambridge University Press.

Bud, R. (1995) 'Science, meaning and myth in the museum', *Public Understanding of Science*, 4, pp. 1–16.

Bukatman, S. (1993) *Terminal Identity: The Virtual Subject in Post-Modern Science Fiction.* Durham, NC, Duke University Press.

Bullas, L. (1979) 'Swinging! That's life for a little miracle', *Manchester Evening News*, 18 October, p. 1.

Bullough, V. (1994) *Science in the Bedroom: A History of Sex Research.* New York, Basic Books.

Burke, M. (1991) 'Monster is born – and it's going on in REAL life, too', *Daily Star*, 9 July, p. 1.

Burnham, J. (1971) *Science in America: Historical Selections.* New York, Holt, Rinehart & Winston.

Burnham, J. (1987) *How Superstition Won and Science Lost: Popularising Science and Health in the United States.* New Brunswick, NJ, Rutgers University Press.

Burroughs, E. (1976) *The Monster Men.* Tandem; first published 1929.

Butler, M. (ed.) (1993) *Frankenstein: The 1818 Text.* William Pickering.

Bynum, W. (1994) *Science and the Practice of Medicine in the Nineteenth Century.* Cambridge University Press.

Calder, N. (1968) 'Depressing futures', *New Statesman*, 26 April, p. 552.

Calder, R. (1968) 'Where are we going?', *New Scientist*, 25 April, p. 190.

Calder, R. (1961) *The Life Savers.* Pan Books; first published 1953.

Čapek, K. (1923) 'The meaning of *R.U.R*', *Saturday Review*, 21 July, p. 79.

Čapek, K. (1962) *R.U.R. and The Insect Play.* Oxford University Press.

Cardwell, D. (1972) *The Organisation of Science in England.* Heinemann.

Carlson, E. (1967) 'The legacy of Hermann Joseph Müller, 1890–1967', *Canadian Journal of Genetics and Cytology*, 9, pp. 436–48.

Carlson, E. (1971) 'An unacknowledged founder of molecular biology: H.J. Müller's contributions to gene theory', *Journal of the History of Biology*, 4, pp. 149–70.

Carlson, E. (1974) 'The Drosophila Group: The Transition from the Mendelian Unit to the Individual Gene', *Journal of the History of Biology*, 7, p. 31–48.

Carlson, E. (1981) *Genes, Radiation and Society: The Life and Work of H.J. Müller.* New York, Cornell University Press.

Carrel, A. (1948) *Man, the Unknown.* West Drayton, Penguin; first published 1935.

Carrel, A. (1967) 'Suture of blood-vessels and transplantation of organs', in *Nobel Lectures, Physiology or Medicine, 1901–1921.* New York, Elsevier.

Carroll, L. (C.L. Dodgson) (1875) 'Some popular fallacies about vivisection', *Fortnightly Review*, 23, p. 854.

Carter, P. (1977) *The Creation of Tomorrow: Fifty Years of Magazine Science Fiction.* New York, Columbia University Press.

Channell, D. (1991) *The Vital Machine: A Study of Technology and Organic Life.* Oxford University Press.

Charbonneur, L. (1976) *Embryo.* New York, Warner Books.

Chargaff, E. (1973) 'Bitter fruits from the tree of knowledge: remarks on the current revulsion from science', *Perspectives in Biology and Medicine*, 16, pp. 486–502.

Chargaff, E. (1976) 'On the dangers of genetic meddling', *Science*, 192, p. 938.

Chargaff, E. (1978) *Heraclitian Fire.* New York, Rockefeller University Press.

Chartres, J. (1970) 'Funds needed to complete implant baby experiment', *The Times*, 27 February, p. 2.

Chedd, G. (1970) 'Biology stirs its conscience', *New Scientist*, 10 December.

Chedd, G. (1974) 'Boston notebook: whose grave?', *New Scientist*, 11 July, p. 91.

Clareson, T. (1959) 'An annotated checklist of American science-fiction: 1880–1915', *Extrapolation*, 1, pp. 5–19.

Clarke, I. (1976) 'From prophecy to prediction. 13. Science and society: prophecies and predictions, 1840–1940', *Futures*, 8, pp. 350–6.

Clarke, I. (1979) *The Pattern of Expectation, 1644–2001*. Jonathan Cape.

Clarke, R. (1968a) *J.B.S. The Life and Work of J.B.S. Haldane*. Hodder & Stoughton.

Clarke, R. (1968b) *The Huxleys*. Heinemann.

Clute, J. and Nicholls, P. (eds) (1993) *The Encyclopaedia of Science Fiction*. Orbit.

Cohen, J. (1966) *Human Robots in Myth and Science*. George Allen & Unwin.

Cohen, S. (1975) 'The manipulation of genes', *Scientific American*, July.

Cohen, S. (1977) 'Recombinant DNA: fact and fiction', *Science*, 195, p. 655.

Cohen, S. and Young, J. (1973) *The Manufacture of News. Deviance, Social Problems and the Mass Media*. Constable.

Coleman, W. (1971) *Biology in the Nineteenth Century: Problems of Form, Function and Transformation*. New York, John Wiley.

Commission of the European Community (1993) *Biotechnology and Genetic Engineering: What Europeans Think About It in 1993. Survey Conducted in the Context of Eurobarometer 39.1*. Brussels, European Commission.

Conklin, G. (1965) *Science Fiction Adventures in Mutation*. New York, Berkley Books; first published 1955.

Cooper, W. (1978) 'Carbon copy babies', *Cosmopolitan*, October.

Cooter, R. and Pumfrey, S. (1994) 'Separate spheres and public places: reflections on the history of science popularization and science in popular culture', *History of Science*, 32, 237–67.

Corner, G. (1964) *A History of the Rockefeller Institute 1901–53: the Early Years*. New York, Rockefeller Institute Press.

Coupland, D. (1997) 'Clone, Clone on the Range', *Time*, 10 March 1997.

Crick, F. (1957) 'On Protein Synthesis', *Symposia of the Society for Experimental Biology*, 12, pp. 138–63.

Crowther, J. (1970) *Fifty Years with Science*. Barrie & Jenkins.

Culliton, B. (1978a) 'Cloning caper makes it to the halls of Congress', *Science*, 200, p. 1250.

Culliton, B. (1978b) 'Scientists dispute book's claim that human clone has been born', *Science*, 199, pp. 1314–16.

Davis, B. (1970) 'Prospects for genetic intervention in man', *Science*, 170, pp. 1279–83.

Davis, B. (1976) 'Novel pressures on the advance of science', *Annals of the New York Academy of Science*, 265, pp. 193–205.

Davis, B. (1991) *The Genetic Revolution: Scientific Prospects and Public Perceptions*. Baltimore, MD, Johns Hopkins University Press.

Davis, K. (1966) 'Sociological aspects of genetic control', in Roslansky, 1966.

Davy, J. (1966) 'New Einsteins from "cuttings"', *Observer*, 13 November, p. 1.

De Camp, L. and Clareson, T. (1978) 'The scientist', in Warrick et al., 1978.

De Kruif, P. (1923) 'Jacques Loeb, the mechanist', *Harper's*, 146, pp. 182–98.

De Kruif, P. (1940) *Microbe Hunters*. New York, Pocket Books; first published 1923.

Dolbear, A. (1905) 'The science problems of the twentieth century', *Popular Science Monthly*, 67, pp. 237–51.

Dowling, D. (1986) 'The atomic scientist: machine or moralist?', *Science Fiction Studies*, 13, pp. 139–47.

Dowse, R. and Palmer, D. (1963) 'Introduction', in Shelley, 1963.

Doyle, Sir A. (1890) 'The Physiologist's Wife', *Blackwood's Edinburgh Magazine*, 148, pp. 339–51.

Doyle, C. (1974) '"Test-tube baby" unit for childless women', *Observer*, 21 July, p. 3.

Dronamraju, K. (ed.) (1969) *Haldane and Modern Biology*. Baltimore, MD, Johns Hopkins University Press.

Dronamraju, K. (ed.) (1995) *Haldane's 'Daedalus' Revisited*. Oxford University Press.

Dubos, R. (1951) *Louis Pasteur: Freelance of Science*. Gollancz.

Dubos, R. (1976) *The Professor, The Institute, and DNA: Oswald T. Avery, his Life and Scientific Achievements*. New York, Rockefeller University Press.

Duffus, F. (1924) 'Jacques Loeb, mechanist', *Century Illustrated Monthly Magazine*, 108, pp. 374–83.

Duncan, R. (1912) 'Some unsolved problems in science', *Harper's*, 125, pp. 28–36.

Durant, J. (ed.) (1992) *Biotechnology in Public: a Review of Recent Research*. London, Science Museum.

Durant, J., Evans, G. and Thomas, G. (1992) 'Public understanding of science in Britain: the role of medicine in the popular representation of science', *Public Understanding of Science*, 1, pp. 161–82.

Durden-Smith, J. (1979) 'Rorvik's baby', *Radio Times*, 223, p. 19.

Dyson, F. (1979) *Disturbing the Universe*. New York, Harper & Row.

Easton, T. (1978) 'The reference library', *Analog*, October, pp. 167–72.

Ebon, M. (1978) *The Cloning of Man. A Brave new Hope – or Horror?* New York, Signet.

Edwards, R. (1969) 'Early stages of fertilisation in vitro of human oocytes matured in vitro', *Nature*, 221, p. 632.

Edwards, R. (1989) *Life Before Birth: Reflections on the Embryo Debate*. Hutchinson.

Edwards, R. and Sharpe, D. (1971) 'Social values and research in human embryology', *Nature*, 231, pp. 87–90.

Edwards, R. and Steptoe, P. (1980) *A Matter of Life. The Story of a Medical Breakthrough*. Hutchinson.

Edwards, W. and Edwards, P. (1974) *Alexis Carel, Visionary Surgeon*. Springfield, IL, Charles C. Thomas.

Eggington, J. (1978) 'American worries hold up genetic research', *Observer*, 23 July.

Einseidel, E. (1992) 'Framing science and technology in the Canadian press', *Public Understanding of Science*, 1, pp. 89–102.

Eisenberg, L. (1977) 'The social imperatives of medical research', *Science*, 198, p. 1105.

Eisendrath, C. (1979) 'The press as guilty bystander', pp. 279–99 in Jackson and Stich 1979.

Ellegard, A. (1958) 'Darwin and the general reader. The reception of Darwin's theory of evolution in the British periodical press, 1859–72', *Goteborg's Universitets Arsskrift*, 64.

Ellison, D. (1978) *The Biotechnical Fix*. New York, Greenwood Press.

Elmer-Dewitt, P. (1993) 'Cloning: where do we draw the line?', *Time*, 8 November, pp. 62–8.

Engel, L. (1962) 'The race to create life', *Harper's*, October, pp. 39–45.

Etzioni, A. (1973) *Genetic Fix*. New York, Collier Macmillan.

Etzioni, A. and Nunn, C. (1976) 'The public appreciation of science in contemporary America', pp. 229–43 in Holton and Blanpied, 1976.

Evans, H. and Evans, D. (1971) *Beyond the Gaslight. Science in Popular Fiction 1885–1905.* Frederick Muller.

Ezrahi, Y. (1971) 'The political resources of American science', *Science Studies*, 1.

Fairclough, P. (ed.) (1968) *Three Gothic Novels.* Harmondsworth, Penguin.

Fara, P. (1996) *Sympathetic Attractions: Magnetic Practices, Beliefs and Symbolism in Eighteenth-Century England.* Princeton, NJ, Princeton University Press.

Farago, P. (1975) *Science and the Media.* Oxford University Press.

Farley, J. (1977) *The Spontaneous Generation Controversy: From Descartes to Oparin.* Baltimore, MD, Johns Hopkins University Press.

Farr, R.M. (1993) 'Common sense, science and social representations', *Public Understanding of Science*, 2, pp. 189–204.

Farrell, L. (1970) 'The origins and growth of the English eugenics movement 1865–1925'. Unpublished PhD thesis, Indiana University, Bloomington.

Flagg, F. (1927) 'The Machine Man of Ardathia', *Amazing Stories*, November, pp. 798–804.

Fleck, J. (1984) 'Artificial intelligence and industrial robots: an automatic end for utopian thought?', pp. 189–231 in M. Mendelsohn and H. Novotny (eds) *Nineteen Eighty-Four: Science between Utopia and Dystopia* (Sociology of the Sciences, Vol. 8). Dordrecht, Reidel.

Fleischer, L. (1994) *Mary Shelley's 'Frankenstein'.* Pan Books.

Fleming, D. (1964) 'Introduction', in Loeb, 1964.

Fleming, D. (1969) 'On living in a biological revolution', *Atlantic Monthly*, 223, pp. 64–70.

Fletcher, J. (1974) *The Ethics of Genetic Control: Ending Reproductive Roulette.* New York, Doubleday.

Flexner, S. (1927) 'Jacques Loeb and his period', *Science*, 66, p. 333.

Florescu, R. (1977) *In Search of Frankenstein.* New English Library.

Forry, S. (1990) *Hideous Progenies: Dramatizations of 'Frankenstein' from Mary Shelley to the Present.* University of Pennsylvania Press.

Fournier, E. (1928) 'Can life riddle be solved?', *Daily Mirror*, 14 September.

Fox, R. and Swazey, J. (1974) *The Courage to Fail: a Social View of Organ Transplants and Dialysis.* Chicago, University of Chicago Press.

Fraenkel-Conrat, H. (1956) 'Rebuilding a virus', *Scientific American*, June, pp. 42–7.

Fraenkel-Conrat (1975) 'Taboo research', *Nature*, 254, p. 12.

Francouer, R. (1972) *Eve's New Rib: Twenty Faces of Sex, Marriage and the Family.* McGibbon & Kee.

Francouer, R. (1973) *Utopian Motherhood.* George Allen & Unwin.

Frankel, C. (1973) 'The nature and sources of irrationalism', *Science*, 180, pp. 927–30.

Frankel, C. (1974) 'The specter of eugenics', *Commentary*, March.

Franklin, H. (1978) *Future Perfect. American Science Fiction of the Nineteenth Century.* Oxford University Press.

Frayling, C. (1992) *Vampyres: Lord Byron to Count Dracula.* Faber & Faber.

French, R. (1975) *Antivivisection and Medical Science in Victorian Society.* Princeton, NJ, Princeton University Press.

Friday, J. (1974) 'A microscopic incident in a monumental struggle: Huxley and antibiosis in 1875', *British Journal for the History of Science*, 7, pp. 61–71.

Frude, N. (1984) *The Robot Heritage.* Century.

Fruton, J. (1972) *Molecules and Life.* New York, Wiley.

Fuller, W. (ed.) (1971) *The Social Impact of Modern Biology.* Routledge & Kegan Paul.

Gabor, D. (1964) *Inventing the Future.* New York, Knopf.

Gale, G. (1978) 'Mixed blessings for Britain's brave new babies', *Daily Express*, 27 July.

Galton, L. (1962) 'Science stands at awesome thresholds', *New York Times*, 2 December, p. vi.

Gaylin, W. (1972) 'The Frankenstein myth becomes a reality', *New York Times Magazine*, 5 March.

Geison, G. (1969) 'The protoplasmic theory of life and the vitalist–mechanist debate', *Isis*, 60, p. 273–92.

Geison, G. (1978) *Michael Foster and the Cambridge School of Physiology: The Scientific Enterprise in Late Victorian Society*. Princeton, NJ, Princeton University Press.

Gibbs, P. (1912) 'The chemistry of life', *The Graphic*, 14 September, p. 406.

Giddens, A. (1990) *The Experience of Modernity*. Cambridge, Polity Press.

Gilbert, S. (1995) 'Resurrecting the body: has postmodernism had any effect on biology?', *Science in Context*, 8, pp. 563–77.

Gilbert, S. and Gubar, S. (1979) *The Madwoman in the Attic*. New Haven, CT, Yale University Press.

Glass, B. (1960) 'The academic scientist, 1940–1960', *Science*, 132, pp. 598–603.

Glass, B. (1963) 'Revolution in biology', in H. Deason, *A Guide to Science Reading*. New York, New American Library.

Glut, D. (1973) *The Frankenstein Legend*. Methuen.

Glut, D. (1984) *The Frankenstein Catalogue*. Jefferson, NC, McFarland.

Goethe, J. (1949,1959) *Faust, Parts One and Two*, trans. Philip Wayne. Harmondsworth, Penguin.

Golden, F. (1978) 'The first test-tube baby', *Time*, 31 July, pp. 46–52.

Goldman, S. (1989) 'Images of technology in popular films: discussion and filmography', *Science, Technology and Human Values*, 14, pp. 275–301.

Goodell, R. (1977) *The Visible Scientists*. Boston, MA, Little, Brown.

Goodfield, J. (1974) 'Changing strategies: a comparison of reductionist attitudes in biological research in the nineteenth and twentieth centuries', pp. 65–87. In Ayala and Dobzhansky, 1974.

Goodfield, J. (1977) *Playing God: Genetic Engineering and the Manipulation of Life*. Hutchinson.

Gore, R. and Dale, B. (1976) 'The new biology', *National Geographic*, September, pp. 355–409.

Gould, S. (1993) 'Dinomania', *New York Review of Books*, 12 August, pp. 51–6.

Graham, L. (1978) 'Concerns about science and attempts to regulate inquiry', *Daedalus*, Spring, pp. 1–22.

Grant, M. (1995) 'James Whale's *Frankenstein*: the horror film and the symbolic biology of the cinematic monster', Chapter 6 in Bann, 1995.

Graubard, M. (1967) 'The Frankenstein syndrome: man's ambivalent attitude to knowledge and power', *Perspectives in Biology and Medicine*, Spring, pp. 418–45.

Gray, G. (1937) 'Where life begins: search among the gene, the virus and the enzyme', *Harper's*, February.

Greenberg, D. (1967) 'The synthesis of DNA: how they spread the good news', *Science*, 158, pp. 1548, 1551.

Greenberg, D. (1969) *The Politics of American Science*. Harmondsworth, Penguin.

Grobstein, C. (1979a) *A Double Image of the Double Helix: The Recombinant DNA Debate*. W.H. Freeman.

Grobstein, C. (1979b) 'External human fertilization', *Scientific American*, June, pp. 33–43.

Grossman, E. (1971) 'The obsolescent mother (a scenario)', *Atlantic*, May.

Grosvenor, R. (1980a) 'How do you tell a child like Louise where babies come from?', *Daily Mirror*, 3 June, pp. 8–9.

Grosvenor, R. (1980b) 'Miracle baby Louise', *Daily Mirror*, 2 June, pp. 16–17.

Gruenberg, A. (1911a) 'The creation of "artificial life", I. The making of living matter from non-living', *Scientific American*, 105, p. 231.

Gruenberg, B. (1911b) 'The creation of "artificial life". 2. Making the non-living do the work of the living', *Scientific American*, 105, p. 302.

Gussin, A. (1963) 'Jacques Loeb: the man and his tropism theory of animal conduct', *Journal of the History of Medicine*, 18, pp. 321–36.

Gwynne, P. (1978) 'All about that baby', *Newsweek*, 7 August.

Hagar, W. (1982) *Terminal Visions: The Literature of Last Things*. Bloomington, Indiana University Press.

Haining, P. (ed.) (1977) *The Frankenstein File*. New English Library.

Haining, P. (ed.) (1994) *The Frankenstein Omnibus*. Orion Books.

Halberstam, J. (1995) *Skin Shows: Gothic Horror and the Technology of Monsters*. Durham, NC, Duke University Press.

Haldane, J.B.S. (1924) *Daedalus or Science and the Future*. Kegan Paul.

Haldane, J.B.S. (1976) *The Man with Two Memories*. Merlin Press.

Hall, S. and Gieben, B. (eds) (1992) *Formations of Modernity*. Cambridge, Polity Press.

Hall, T. (1969) *Ideas of Life and Matter*, 2 vols. Chicago, University of Chicago Press.

Haller, M. (1963) *Eugenics. Hereditarian Attitudes in American Thought*. New Brunswick, NJ, Rutgers University Press.

Hamilton, E. (1936) 'Master of the Genes', *Wonder Stories*, November.

Hammond, R. (1986) *The Modern Frankenstein: Fiction Becomes Fact*. Poole, Blandford Press.

Hampson, J. (1979) 'Animal welfare – a century of conflict', *New Scientist*, 25 October.

Handlin, O. (1972) 'Ambivalence in the popular response to science', in Barnes, 1972; first published as 'Science and technology in popular culture', in G. Holton (ed.) *Science and Culture*. Boston, MA, Houghton Mifflin.

Hanmer, J. (1980) 'Reproductive engineering: the final solution?', *New Society*, 24 July, pp. 163–4.

Harkins, W. (1962) *Karel Čapek*. New York, Columbia University Press.

Harris, C. (1929) 'The Evolutionary Monstrosity', *Amazing Stories Quarterly*, Winter.

Harris, P. (1978) 'It's the all-British miracle', *Manchester Evening News*, 26 July.

Harrison, R. (1907) 'The growth of living nerve cells in vitro', *Proceedings of the Society for Experimental Biology and Medicine*, 4.

Hart, A. (1988) *Making the Real World*. Cambridge University Press.

Hawkins, W. (1934) 'The Regenerative Wonder', *Amazing Stories*, February.

Haynes, R. (1994) *From Faust to Strangelove: Representations of the Scientist in Western Literature*. Baltimore, MD, Johns Hopkins University Press.

Haynes, R. (1995) 'Frankenstein: the scientist we love to hate', *Public Understanding of Science*, 4, pp. 435–44.

Heckstall-Smith, H. (1958) *Atomic Radiation Dangers – and What They Mean To You*. J.M. Dent.

Heinlein, R. (1953) 'Jerry Was a Man', in *Assignment in Eternity*, Berkeley, CA, Fantasy Press; first published 1947.

Hendrick, B. (1913) 'On the trail of immortality', *World's Work*, March.

Herbert, F. (1966) *The Eyes of Heisenberg.* New York, Berkley Books.

Hill, A. (1962) *The Ethical Dilemma of Science and Other Writings.* Science Book Guild.

Hill, G. (1992) *Illuminating Shadows: The Mythic Power of Film.* Shambhala.

Hillegas, M. (1967) *The Future as Nightmare. H.G. Wells and the Anti-Utopians.* Oxford University Press.

Hindle, M. (ed.) (1985) *Frankenstein.* Harmondsworth, Penguin.

Hobsbawm, E. (1964) *The Age of Revolution.* New York, New American Library.

Hobsbawm E. (1994) *Age of Extremes: The Short Twentieth Century, 1914–1991.* Michael Joseph.

Hodgkinson, P. (1978) 'The battles – and the breakthrough', *Daily Mail,* 27 July.

Hofstadter, R. (1959) *Social Darwinism in American Thought.* New York, George Braziller.

Holton, G. (1976) 'Scientific optimism and social concerns', *Annals of the New York Academy of Sciences,* 265, pp. 82–101.

Holton, G. (1993) 'Can science be at the centre of modern culture?', *Public Understanding of Science,* 2, pp. 291–306.

Holton, G. and Blanpied, W. (eds) (1976) *Science and its Public: The Changing Relationship. Boston Studies in the Philosophy of Science,* XXXIII, Boston, D. Reidel.

Howard, T. and Rifkin, J. (1977) *Who Should Play God? The Artificial Creation of Life and What It Means for the Future of the Human Race.* New York, Delacorte Press.

Huet, M.-H. (1993) *Monstrous Imagination.* Cambridge, MA, Harvard University Press.

Hunter, J.P. (ed.) (1996) *Frankenstein.* Norton.

Hutchings, P. (1993) *Hammer and Beyond: the British Horror Film.* Manchester, Manchester University Press.

Hutton, R. (1978) *Bio-Revolution. DNA and the Ethics of Man-made Life.* New York, Mentor Books.

Huxley, A. (1932) *Brave New World.* Chatto & Windus.

Huxley, A. (1950) *Brave New World,* Harmondsworth, Penguin.

Huxley, J. (1926) 'The Tissue Culture King', *Yale Review,* 15, p. 479. Also *Cornhill Magazine,* 60, p. 422. Reprinted (1927) *Amazing Stories,* August. Reprinted in Conklin, G. (ed.) (1962) *Great Science Fiction by Scientists.* Collier Macmillan.

Huxley, J. (1937) *Essays in Popular Science.* Harmondsworth, Penguin; first published 1926.

Huxley, J. (1939) *Essays of a Biologist.* Harmondsworth, Penguin; first published 1923.

Huxley, J. (1964) *Essays of a Humanist.* Harper.

Huxley, J. (1970) *Memories,* Vol. 1. George Allen & Unwin.

Huxley, T. (1862) *On Our Knowledge of the Causes of the Phenomena of Organic Nature* (Six lectures to working men). Robert Hardwicke.

Huxley, T. (1869) 'On the physical basis of life', *Fortnightly Review,* 5, pp. 129–45.

Huxley, T. (1871) *Biogenesis and Abiogenesis, Report of the British Association for the Advancement of Science, 1870.* BAAS.

Jackson, D. and Stich, S. (eds) (1979) *The Recombinant DNA Debate.* Englewood Cliffs, NJ, Prentice-Hall.

Jacob, F. (1982) *The Logic of Life. A History of Heredity.* New York, Pantheon Books.

James, F. and Field, J. (1994) 'Frankenstein and the spark of being', *History Today,* 449, pp. 47–53.

Jennings, A. (1978) 'Hunting the test-tube baby mum', *New Manchester Review,* 28 July.

Jaynes, G. (1995) 'Meet Mister Wizard – Michael Crichton', *Time,* 2 October, p. 69.

Jessel, S. (1969) 'Views on experiments divided', *The Times,* 15 February.

Jolly, W. (1974) *Sir Oliver Lodge: Psychical Researcher and Scientist.* Constable.

Jones, G., Connell, I. and Meadows, J. (1978) *The Presentation of Science by the Media.* Leicester, Primary Communications Research Centre.

Jones, H. (1971) *The Age of Energy: Varieties of American Experience, 1865–1915.* New York, Viking Press.

Jones, S. (1994) *The Illustrated 'Frankenstein' Movie Guide.* Titan Books.

Jordanova, L. (1989) *Sexual Visions: Images of Gender in Science and Medicine between the Eighteenth and Twentieth Centuries.* Brighton, Harvester Wheatsheaf.

Judson, H. (1979) *The Eighth Day of Creation. Makers of the Revolution in Biology.* New York, Simon & Schuster.

Jukes, T. (1976) 'Wildcat story', *Nature*, 262, p. 736.

Jukes, T. (1977) 'Monster debate', *Nature*, 266, p. 215.

Kaempffert, W. (1935) 'The year in science: progress is achieved in many fields', *New York Times*, 29 December, p. iv.

Kaempffert, W. (1936) *Science, Today and Tomorrow.* New York, Viking Press.

Kass, L. (1971a) 'Babies by means of *in vitro* fertilization: unethical experiments on the unborn?', *New England Journal of Medicine*, 285, p. 1174.

Kass, L. (1971b) 'The new biology: what price relieving man's estate?', *Science*, 174, pp. 779–88.

Kass, L. (1972) 'Making babies: the new biology and the "old" morality', *Public Interest*, 26, p. 18.

Kay, L. (1993) *The Molecular Vision of Life: Caltech, The Rockefeller Foundation and the Rise of the New Biology.* Oxford University Press.

Keller, E. (1985) *Reflections on Gender and Science.* New Haven, CT, Yale University Press.

Keller, E. (1992) *Secrets of Life, Secrets of Death: Essays on Language, Gender and Science.* Routledge, 1992.

Kendrew, J. (1966) *The Thread of Life.* G. Bell & Sons.

Ketterer, D. (1978) 'Mary Shelley and science fiction – a select bibliography', *Science Fiction Studies*, 5, pp. 172–8.

Kevles, D. (1986) *In the Name of Eugenics: Genetics and the Uses of Human Heredity.* Harmondsworth, Penguin.

Kevles, D. (1992) 'Huxley and the popularization of science', pp. 238–51 in Waters and Van Helden, 1992.

Kimbrell, A. (1993) *The Human Body Shop: The Engineering and Marketing of Life.* HarperCollins.

King-Hele, D. (1963) *Erasmus Darwin.* Macmillan.

Kleier, J. (1928) 'The Head.' *Amazing Stories*, August.

Kline, O. (1926) 'The Malignant Entity', *Amazing Stories*, June.

Klingender, F. (1972) *Art and the Industrial Revolution.* Granada.

Kluger, J. (1997) 'Will We Follow the Sheep?' *Time*, 10 March, p. 49.

Knoepfelmacher, H. and Tennyson, G. (eds) (1978) *Nature and the Victorian Imagination.* Berkeley, University of California Press.

Kohler, R. (1972) 'The reception of Eduard Buchner's discovery of cell-free fermentation', *Journal of the History of Biology*, 5, pp. 327–53.

Kohler, R. (1976) 'The management of science: the experience of Warren Weaver and the Rockefeller Foundation programme in molecular biology', *Minerva*, 14, pp. 279–306.

Kohler, R. (1978) Fletcher, Hopkins, and the Dunn Institute of Biochemistry 'A case study in the patronage of science', *Isis*, 69, pp. 331–55.

Krieghbaum, H. (1968) *Science and the Mass Media.* University of London Press.

Krimsky, S. (1982) *Genetic Alchemy: The Social History of the Recombinant DNA Debate.* Cambridge, MA, MIT Press.

Krimsky, S. (1992) 'Regulating recombinant DNA research and its applications', Chapter 12 in D. Nelkin, (ed.) *Controversy: Politics of Technical Decisions*, 3rd edn. Sage.

Ladenis, N. (1996) 'Soya beans go bananas', *Guardian*, 5 December, p. 21.

Lady, S. and Darabont, F. (1994) 'The screenplay', in K. Branagh, *Mary Shelley's 'Frankenstein'.* Pan Books.

La Follette, M. (1990) *Making Science Our Own: Public Images of Science, 1910–1955.* Chicago, University of Chicago Press.

La Mettrie, J. (1996) *Machine Man and Other Writings.* Cambridge University Press.

Lander, E. (1996) 'The new genomics: global views of biology', *Science*, 274, pp. 536–39.

Lankester, E. (1913) *More Science from an Easy Chair.* Methuen.

Lansbury, C. (1985) *The Old Brown Dog: Women, Workers and Vivisection in Edwardian England.* Madison, University of Wisconsin Press.

Lappe, M. (1972) 'Risk taking for the unborn', *Hastings Center Report*, 2, p. 1.

Lappe, M. (1979) *Genetic Politics: The Limits of Biological Control.* New York, Simon & Schuster.

Lappe, M. and Morison, A. (eds) (1976) 'Ethical and scientific issues posed by human uses of molecular genetics', *Annals of the New York Academy of Sciences*, 265.

Laurence, W. (1935) 'Carrel, Lindbergh, develop device to keep organs alive outside body', *New York Times*, 21 June.

Lavelley, A. (1979) 'The Stage and Film Children of Frankenstein', Chapter xi, pp. 243–89 of G. Levine and U.C. Knoepflmacher (eds), *The Endurance of Frankenstein – Essays on Mary Shelley's Novel*, Berkeley, University of California Press.

Leach, E. (1968) *A Runaway World?* Oxford University Press.

Leach, G. (1970) 'The test-tube fantasy', *Observer*, 1 March 1970.

Leach, G. (1972) *The Biocrats: Implications of Medical Progress.* Harmondsworth, Penguin.

Lederburg, J. (1970) 'Genetic engineering and the amelioration of genetic defect', *BioScience*, 20, pp. 1307–10.

Lehman, S. (1972) 'The motherless child in science fiction: *Frankenstein* and *Moreau*', *Science Fiction Studies*, 19, pp. 49–57.

Leiss, W. (1972) *The Domination of Nature.* New York, George Braziller.

LeMahieu, D. (1992) 'The ambiguity of popularization', pp. 252–6 in Waters and Van Helden, 1992.

Lemonick, M. (1993) 'Cloning classics', *Time*, 8 November 1993, p. 68.

Le Palourel, H. (1970) 'Test-tube life' (letter), *The Times*, 25 February, p. 9.

Lessing, L. (1967) *DNA: at the Core of Life Itself.* New York, Collier Macmillan.

Levene, P. (1924) 'Jacques Loeb – the man', *Science*, 59, pp. 427–8.

Levine, G. and Knoepflmacher, U. (1982) *The Endurance of 'Frankenstein': Essays on Mary Shelley's Novel.* Berkeley, University of California Press.

Lewin, R. (1977a) 'Demonstration disrupts genetic engineering forum', *New Scientist*, 17 March, p. 628.

Lewin, R. (1977b) 'The mayor and the monster', *New Scientist*, 6 October, p. 16.

Lewin, R. (1979) 'A question of confidence', *New Scientist*, 12 April, p. 98.

Lewis, S. (1925) *Martin Arrowsmith.* Jonathan Cape.

Lewontin, R. (1968) 'Essay review of *Phage and the Origins of Molecular Biology*', *Journal of the History of Biology*, 1, pp. 155–61.

Lewontin, R. (1993) *The Dream of the Human Genome,* in Lewontin, *The Doctrine of DNA.* Penguin. Originally, 1992 *New York Review of Books,* 28 May.

Lindee, S. (1994) *Suffering Made Real: American Science and the Survivors at Hiroshima.* Chicago, University of Chicago Press.

Lipking, L. (1996) 'Frankenstein, the true story: or, Rousseau judges Jean-Jacques', pp. 313–31 in Hunter, 1996.

Livingston, A. (1927) 'The American Association for the Advancement of Science – newspaper reports on the meetings', *Science,* 66, p. 369.

Loeb, J. (1899) 'On the nature of the process of fertilization and the artificial production of normal larvae plutei from the unfertilized eggs of the sea urchin', *American Journal of Physiology,* 3, pp. 135–8.

Loeb, J. (1900) 'On artificial parthenogenesis in sea-urchins,' *Science,* 11, pp. 612–14.

Loeb, J. (1904) 'The recent development of biology,' *Science,* 20, pp. 777–86.

Loeb, J. (1905) *Studies in General Physiology.* (Decennial Publications of the University of Chicago, Second Series, Vol. 15). Chicago, T. Fisher Unwin.

Loeb, J. (1913) *Artificial Parthenogenesis and Fertilization.* Chicago, University of Chicago Press.

Loeb, J. (1964) *The Mechanistic Conception of Life,* with an historical introduction by Donald Fleming. Cambridge, MA, Harvard University Press; first published 1912.

Longo, C. (1976) *The Last Gene.* Canoga Park, CA, Major Books.

Lowe-Evans, M. (1993) *Frankenstein: Mary Shelley's Wedding Guest.* New York, Twayne.

Lubow, A. (1977) 'Playing God with DNA', *New Times,* 7 January, p. 48. Reproduced in J. Watson, and J. Tooze, *The DNA Story,* pp. 118–26.

Ludmerer, K. (1972) *Genetics and American Society: A Historical Appraisal.* Baltimore, MD, Johns Hopkins University Press.

McConnell, F. (1981) 'Evolutionary fables: *The Time Machine* and *The Island of Dr Moreau*', Chapter 3, pp. 69–105 in *The Science Fiction of H.G. Wells,* Oxford University Press.

McCorduck, P. (1979) *Machines Who Think.* W.H. Freeman.

McGregor, A. (1993) 'The Crichton Factor', *Time Out,* 7 April, p. 21.

McIntosh, J. (1961) *The Fittest.* Corgi Books.

Mackenzie, D. (1976) 'Eugenics in Britain', *Social Studies of Science,* 6, pp. 499–532.

McKinnell, R. (1979) *Cloning: A Biologist Reports.* Minneapolis, University of Minnesota Press.

McLeod, J., Kosicki, G. and Zhingdang, P. (1991) 'On understanding and misunderstanding media effects', pp. 235–66 in J. Curran and M. Gurevitch (eds) *Mass Media and Society.* Edward Arnold.

Macmillan, J. (1969) 'Test-tube babies' (letter), *The Times,* 15 February.

McNeil, M. (1987) *Under the Banner of Science: Erasmus Darwin and His Age.* Manchester, Manchester University Press.

Maddox, J. (1969) 'Test-tube babies – the hazards and the hopes', *The Times,* 26 February, p. 11.

Maehle, A. and Trohler, U. (1987) 'Animal experimentation from antiquity to the end of the eighteenth century: attitudes and arguments', pp. 140–7 in Rupke, 1987.

Maienschein, J. (1991) *Transforming Traditions in American Biology, 1880–1915.* Baltimore, MD, Johns Hopkins University Press.

Margerison, T. (1962) 'Architects of life', *Sunday Times Colour Magazine,* 9 December.

Marshall, E. (1979) 'Public attitudes to technological progress', *Science,* 205, pp. 281–5.

Marshall, T. (1995) *Murdering to Dissect. Grave-robbing, 'Frankenstein', and the Anatomy*

Literature. Manchester, Manchester University Press.

Martin, H.-J. (1981) 'Printing', Chapter 6 (pp. 127–50) in R. Williams (ed.) *Contact: Human Communication and its History.* Thames & Hudson.

Martin, S. and Tait, J. (1993) *Release of Genetically Modified Organisms: Public Attitudes and Understanding.* Milton Keynes, Centre for Technology Strategy, Open University.

Marx, J. (1973) 'In vitro fertilization of human eggs: bioethical and legal considerations', *Science*, 182, p. 812.

Marx, L. (1964) *The Machine in the Garden: Technology and the Pastoral Ideal in America.* Oxford University Press.

May, R. (1977) 'The recombinant DNA debate', *Science*, 198, pp. 1145–6.

Mazlish, B. (1993) *The Fourth Discontinuity: The Co-Evolution of Humans and Machines.* New Haven, CT, Yale University Press.

Medawar, P. (1977) 'Fear and DNA', *New York Review of Books*, 29 October.

Mellor, A. (1988) *Mary Shelley: Her Life, Her Fiction, Her Monsters.* Routledge.

Mendelsohn, E. (1979) 'Frankenstein at Harvard: the public politics of recombinant DNA research', in E. Mendelsohn, P. Weingart and D. Nelkin (eds) *The Social Assessment of Science: Issues and Perspectives.* Bielefeld, University of Bielefeld.

Mendelsohn, E. (1984) '"Frankenstein at Harvard": the public politics of recombinant DNA research', Chapter 16 in E. Mendelsohn (ed.) *Transformation and Tradition in the Sciences: Essays in Honour of I. Bernard Cohen.* Cambridge University Press.

Mertens, T. (ed.) (1975) *Human Genetics: Readings on the Implications of Genetic Engineering.* Chichester, John Wiley.

Miller, W. (1974) 'Reproduction, technology and the behavioural sciences', *Science*, 183, p. 149.

Millhauser, M. (1973) 'Dr Newton and Mr Hyde: scientists in fiction from Swift to Stevenson', *Nineteenth Century Fiction*, 28, pp. 287–304.

Moores, S. (1993) *Interpreting Audiences: The Ethnography of Media Consumption.* Sage.

Morgan, H. (1971) *Unity and Culture: American History 1877–1900.* Harmondsworth, Penguin.

Morison, R. (1968) 'Man changing man', *New York Times Book Review*, 15 September, p. 3.

Moskowitz, S. (ed.) (1973) *Science Fiction by Gaslight: A History and Anthology of Science Fiction in the Popular Magazines, 1891–1911.* Westport, CT, Hyperion Press.

Motulsky, A. (1974) 'Brave new world?', *Science*, 185, pp. 653–63.

Mulkay, M. (1993) 'Rhetorics of hope and fear in the Great Embryo Debate', *Social Studies of Science*, 23, pp. 721–42.

Mulkay, M. (1996) 'Frankenstein and the debate over embryo research', *Science, Technology and Human Values*, 21, pp. 157–76.

Mullen, R. (1967) 'H.G. Wells and Victor Rousseau Emanuel, *When the Sleeper Wakes* and *The Messiah of the Cylinder*', *Extrapolation*, 8, pp. 31–63.

Müller, H. (1926) 'The gene as the basis of life'. Paper read to the International Congress of Plant Science. Cited in Müller, 1962.

Müller, H. (1939) Manifesto, inserted in an unsigned article: 'Men and mice at Edinburgh: reports from the Genetics Congress', *Journal of Heredity*, 30, p. 371.

Müller, H. (1948) 'Time-bombing our descendants', *American Weekly*, 3 January.

Müller, H. (1956) 'Race poisoning by radiation', *Saturday Review*, 9 June, pp. 9–11, 37–59.

Müller, H. (1961a) 'Genetic nucleic acid: the key material in the origin of life', *Perspectives in Biology and Medicine*, 5, pp. 1–23.

Müller, H. (1961b) 'Human evolution by voluntary choice of germ plasm', *Science*, 134, p. 646.

Müller, H. (1962) *Studies in Genetics: Collected Papers of H. Müller*. Bloomington, Indiana University Press.

Müller, H. (1964a) 'Perspectives for the life sciences', *Bulletin of the Atomic Scientists*, 20, pp. 3–7.

Müller, H. (1964b) 'Production of mutations' (Nobel lecture), 12 December 1946, p. 171 in *Nobel Lectures, Physiology or Medicine, 1942–62*. Amsterdam, Elsevier.

Munnings, B. (1979) 'Horror on Animal Farm – our verdict VILE', *Sunday Mirror*, 24 June.

Murray, M. (1970) 'Test-tube life' (letter), *The Times*, 25 February, p. 9.

Nally, M. (1978) 'Shock horror scoop probe!,' *Observer*, 7 May, p. 3.

Needham, J. (1932) 'Biology and Mr Huxley', *Scrutiny*, May, pp. 76–9. Reprinted in Watt, 1975.

Nelkin, D. (1977a) *Science Textbook Controversies and the Politics of Equal Time*. Cambridge, MA, MIT Press.

Nelkin, D. (1977b) *Technological Decisions and Democracy*. Sage.

Nelkin, D. (1978) 'Threats and Promises – Negotiating the Control of Research', *Daedalus*, Spring, p. 191.

Nelkin, D. (1987) *Selling Science: How the Press Covers Science and Technology*. W.H. Freeman.

Nelkin, D. (1995) 'Forms of intrusion: comparing resistance to information technology and biotechnology in the USA', pp. 379–92 in Bauer, 1995.

Nelkin, D. and Lindee, S. (1995) *The DNA Mystique: The Gene as a Cultural Icon*. W.H. Freeman.

Nelson, G. and Ray, J. (eds) (1974) *Contemporary Readings in Biology*. Chichester, John Wiley.

Newmark, P. (1980) 'Engineered *E. Coli* produce interferon', *Nature*, 283, p. 323.

Nirenberg, M. (1967) 'Will society be prepared?', *Science*, 157, p. 633.

Nitchie, E. (1953) *Mary Shelley: Author of 'Frankenstein'*. New Brunswick, NJ, Rutgers University Press.

O'Brien, R. (1959) 'Coming: man-made men', *Esquire*, April, pp. 95–8.

O'Flinn, P. (1986) 'Production and reproduction: the case of *Frankenstein*', pp. 196–221 in P. Humm, P. Stigant and P. Widdowson (eds) *Popular Fictions: Essays in Literature and History*. Methuen, 1986; first published in *Literature and History*, 9(2), 1983.

Olby, R. (1974a) 'The origins of molecular genetics', *Journal for the History of Biology*, 7, pp. 93–100.

Olby, R. (1974b) *The Path to the Double Helix*. Macmillan.

Olby, R. (1990) 'The molecular revolution in biology', pp. 503–20 in R. Olby, G. Cantor, J. Christie and M. Hodge, *Companion to the History of Modern Science*. Routledge.

Ong, W. (1982) *Orality and Literacy: The Technologizing of the Word*. Methuen.

Osterhout, W. (1928) 'Jacques Loeb', *Journal of General Physiology*, 8, p. 785.

Owen, L. (1979) 'The spectre in the test tube', *Guardian*, 19 February p. 9.

Ozolins, A. (1976) 'Recent work on Mary Shelley and *Frankenstein*', *Science Fiction Studies*, 3, pp. 187–202.

Packard, V. (1978) *The People Shapers*. McDonald & Janes.

Parker, H. (1984) *Biological Themes in Modern Science Fiction*. Ann Arbor, MI, UMI Research Press.

Parkes, A. (1966) *Sex, Science and Society*, Newcastle, Oriel Press.

Parkin, M. (1974) 'Researcher to stop test-tube implants', *Guardian*, 19 July, p. 6.

Parrinder, P. (1972) *H.G. Wells: The Critical Heritage*. Routledge & Kegan Paul.

Parrinder, P. (1995a) 'A sense of dethronement – *The Time Machine* and *The Island of Doctor Moreau*', Chapter 4, pp. 49–64 in *Shadows of the Future: H.G. Wells, Science Fiction and Prophecy*. Liverpool, Liverpool University Press.

Parrinder, P. (1995b) *Shadows of the Future: H.G. Wells, Science Fiction and Prophecy*. Liverpool, Liverpool University Press.

Patten, R. (1992) 'The British context of Huxley's popularization', pp. 257–64 in Waters and Van Helden, 1992.

Paul, D. (1995) *Controlling Human Heredity: 1865 to the Present*. Atlantic Highlands, NJ, Humanities Press.

Pauly, P.J. (1987) *Controlling Life: Jacques Loeb and the Engineering Ideal in Biology*. New York, Oxford University Press.

Pfeiffer, J. (1959) 'To make man better', *New York Times Book Review*, 26 July, p. 21.

Pincher, C. (1970) 'Test-tube babies – soon we'll hardly need people', *Daily Express*, 25 February, p. 8.

Pirie, N. (1966) 'John Burdon Sanderson Haldane, 1892–1964', *Biographical Memoirs of Fellows of the Royal Society*, 12, pp. 219–49.

Pohl, F. and Kornbluth, C. (1952) *Gravy Planet*. Galaxy.

Pollack, R. (1994) *Signs of Life: The Language and Meanings of DNA*. Viking.

Pontecorvo, G. (1968) 'Hermann Joseph Müller, 1890–1967', *Biographical Memoirs of Fellows of the Royal Society*, 14, pp. 349–89.

Poole, J. and Andrews, K. (1972) *The Government of Science in Britain*. Weidenfeld & Nicolson.

Porter, R. (1991) 'History of the body', in P. Burke (ed.) *New Perspectives in Historical Writing*. Cambridge, Polity Press.

Portugal, F. and Cohen, J. (1977) *A Century of DNA: a History of the Discovery of the Structure and Function of the Genetic Substance*. Cambridge, MA, MIT Press.

Potter, D. (1968) 'Biological revolution: but what else?', *The Times*, 27 April, p. 21.

Powledge, T. (1974) 'Dangerous research and public obligation', *New York Times*, 24 August, p. 25.

Praz, M. (1968) 'Introductory Essay', pp. 7–34 in Fairclough, 1968.

Rabinow, P. (1992) 'Artificiality and enlightenment: from sociobiology to biosociality', in J. Crary and S. Kwinter (eds) *Incorporations*. New York, Zone Books.

Rabinow, P. (1996) *Essays on the Anthropology of Reason*. Princeton, NJ, Princeton University Press.

Raknem, I. (1964) *H.G. Wells and his Critics*. Oslo, Universitetsforlaget.

Ramsey, P. (1970) *Fabricated Man: The Ethics of Genetic Control*. New Haven, CT, Yale University Press.

Ramsey, P. (1972) 'Shall we "reproduce"? Part 1. The medical ethics of in vitro fertilization', *Journal of the American Medical Association*, 220, pp. 1346–50; 'Part 2. Rejoinders and future forecasts', ibid., pp. 1480–5.

Ramsey, P. (1975a) *The Ethics of Fetal Research*. New Haven, CT, Yale University Press.

Ramsey, P. (1975b) Genetic engineering', in Mertens, 1975; first published in *Bulletin of the Atomic Scientists*, 1972.

Ratcliff, J. (1937) 'No Father to Guide Them', *Collier's Magazine*, 20 March, p. 19.

Ravetz, J. (1971) *Scientific Knowledge and its Social Problems*. Oxford University Press.

Redfearn, J. (1979) 'Antivivisection demo hits Cambridge', *Nature*, 279, p. 91.

Reich, W. (ed.) (1979) *Encyclopedia of Bioethics*, 4 vols. New York, Free Press.

Reiger, J. (1974) 'Introduction', in M. Shelley, *Frankenstein*. New York, Bobbs-Merrill.

Reilly, P. (1977) *Genetics, Law and Social Policy*, Cambridge, MA, Harvard University Press.

Reingold, N. (1962) 'Jacques Loeb, the scientist, his papers and his era', *Library of Congress Quarterly Journal of Current Acquisitions*, 19, pp. 119–30.

Reinhold, R. (1967) 'Genes are held able to cure diseases', *New York Times*, 22 October.

Reinhold, R. (1968) 'Evolution control: a genetic advance', *New York Times*, 8 September.

Rheinberger, H.-J. (1995) 'Beyond nature and culture: a note on medicine in the age of molecular biology', *Science in Context*, 8, pp. 249–63.

Richards, J. (1979) *Recombinant DNA: Science, Ethics and Politics*. Academic Press.

Richardson, R. (1988) *Death, Dissection and the Destitute*. Routledge.

Rivers, C. (1972) 'Genetic engineering portends a grave new world', *Saturday Review*, 8 April.

Robertson, T. (1926) 'The life and work of a mechanistic philosopher – Jacques Loeb', *Science Progress*, 21, pp. 114–29.

Rock, A. (ed.) (1964) *The Origins and Growth of Biology*. Harmondsworth, Penguin.

Rogers, M. (1975) 'The Pandora's Box congress', *Rolling Stone*, 19 June, p. 36.

Rogers, M. (1977) *Biohazard*. New York, Knopf.

Roll-Hansen, N. (1978) 'Drosophila genetics; a reductionist research programme', *Journal of the History of Biology*, 11, pp. 159–210.

Rollin, B. (1995) *The Frankenstein Syndrome: Ethical and Social Issues in the Genetic Engineering of Animals*. Cambridge University Press.

Romanshyn, R. (1989) *Technology as Symptom and Dream*. Routledge.

Rorvik, D. (1978a) *Brave New Baby: Promise and Peril of the Biological Revolution*. New English Library. First published 1971.

Rorvik, D. (1978b) *In His Image: The Cloning of a Man*. Hamish Hamilton.

Rose, S. and Rose, H. (1970) *Science and Society*. Harmondsworth, Penguin.

Rosenberg, C. (1976) 'Martin Arrowsmith, the scientist as hero', Chapter 7, pp. 123–31 in *No Other Gods: On Science and American Social Thought*. Baltimore, MD, Johns Hopkins University Press.

Rosenberg, R. (1977) 'Of tortoises and men: the development of electro-mechanical organisms and their implication for society', *Artificial Intelligence Society Bulletin*, October, p. 6.

Rosenberg, R. (n. d.) 'The "thinking machine" arrives but only as a child', *Proceedings of the first CSCSI/SCEIO National Conference* (Xerox).

Rosenfeld, A. (1965) 'Will man direct his own evolution?', *Life*, 1 November, pp. 52–8.

Rosenfeld, A. (1972) *The Second Genesis: The Coming Control of Life*. New York, Pyramid Communications.

Roslansky, G. (1966) *Genetics and the Future of Man*. Amsterdam, North Holland.

Rostand, J. (1955) *Life, The Great Adventure: Discussions with Paul Bodin*, trans. A. Brodwick. Hutchinson.

Rostand, J. (1959) *Can Man Be Modified?* trans. J. Griffin. Secker & Warburg.

Roszak, T. (1974) *The Monster and the Titan*. Daedalus 103, 3, p. 31.

Roux, W. (1908) 'The artificial generation of life', *Knowledge and Scientific News*, 29, pp. 484–7.

Rupke, N. (ed.) (1987) *Vivisection in Historical Perspective*. Routledge.

Russell, B. (1924) *Icarus, or the Future of Science*. Kegan Paul.

Russell, B. (1931) *The Scientific Outlook*. George Allen & Unwin.

Russell, B. (1932) 'We don't want to be happy', *New Leader*, 11 March, p. 9. Reprinted in Watt, 1975, pp. 210–11.

Samuelseon, J. (1870) 'The controversy on spontaneous generation: with recent experiments', *Journal of Science*, 7, pp. 484–97.

Sapiro, M. (1964) 'The Faustus Tradition in the Early Science Fiction Story', *Riverside Quarterly*, 1, p. 3.

Sargent, P. (1976) *Cloned Lives*. Greenwich, CT, Fawcett Publications.

Sawday, J. (1995) *The Body Emblazoned: Dissection and the Human Body in Renaissance Culture*. Routledge.

Schachner, N. (1934) 'The 100th Generation', *Astounding*, May.

Schank, R. (1990) *Tell Me a Story: A New Look at Real and Artificial Memory*. New York, Scribner.

Schelde, P. (1993) *Androids, Humanoids and Other Science Fiction Monsters. Science and Soul in Science Fiction Films*. New York, New York University Press.

Schmeck, H.M. (1967) 'Core of virus is made artificially', *New York Times*, 15 December.

Scholes, R. and Rabkin, E. (1977) *Science Fiction: History, Science, Vision*. Oxford University Press.

Schoon, N. (1996) 'Nothing to fear from techno-corn', *Independent*, 11 December, p. 15.

Schuster, A. (1969) 'Human egg fertilizd in test-tube by Britons', *New York Times*, 15 February, p. 31.

Searle, G. (1976) *Eugenics and Politics in Britain, 1900–1914*. Leyden, Noordhoff.

Secord, J. (1989) 'Extraordinary experiment: electricity and the creation of life in Victorian England', Chapter 11 in D. Gooding, T. Pinch and S. Schaffer (eds) *The Uses of Experiment: Studies in the Natural Sciences*. Cambridge University Press.

Serviss, G. (1905) 'Artificial creation of life', *Cosmopolitan*, 39, p. 459.

Seymour-Ure, C. (1977) *Science and Medicine and the Press* (Royal Commission on the Press, Working Paper 3, Studies on the Press), 45–82. HMSO.

Shapin, S. (1990) 'Science and the public', pp. 980–9 in R. Olby et al. *Companion to the History of Modern Science*. Routledge.

Sherrington, C. (1963) *Man on His Nature*. Cambridge University Press.

Shipley, M. (1931) 'Has living matter been produced in the laboratory?' *Scientific American*, 144, p. 18.

Silcock, B. (1970) 'Test-tube babies – the risk', *Sunday Times*, 1 March, p. 11.

Silverberg, R. (ed.), (1977) *Mutants*, Corgi Books.

Simmons, J. (1928) 'The Living Test Tube', *Amazing Stories*, November.

Singer, M. (1977a) 'Scientists and the control of science', *New Scientist*, 16 June.

Singer, M. (1977b) 'The recombinant DNA debate', *Science*, 196, p. 127.

Singer, M. and Soll, D. (1973) 'Guidelines for DNA hybrid molecules', *Science*, 181, p. 1114.

Sinsheimer, R. (1974) 'The brain of Pooh: an essay on the limits of mind', in Nelson and Ray, 1974, pp. 120–38; first published in *Engineering and Science*, January 1970.

Sinsheimer, R. (1975) 'Troubled dawn for genetic engineering', *New Scientist*, 16 October.

Sinsheimer, R. (1977) 'An evolutionary perspective for genetic engineering', *New Scientist*, 20 January.

Sinsheimer, R. (1978) 'The presumptions of science', *Daedalus*, Spring, pp. 223–45.

Skal, D. (1990) *Hollywood Gothic: The Tangled Web of 'Dracula' from Novel to Stage to Screen*. New York, W.W. Norton.

Sklair, L. (1970) *The Sociology of Progress*. Routledge & Kegan Paul.

Small, C. (1972) *Ariel Like a Harpy: Shelley, Mary and 'Frankenstein'*. Victor Gollancz.

Smith, A. (1969) '"Babies in a test-tube" claim by scientists', *Daily Mirror*, 14 February, p. 1.

Smith, A. (1975) *The Human Pedigree: Inheritance and the Genetics of Mankind*. George Allen & Unwin.

Smith, J. (ed.) (1992) *'Frankenstein': Case Studies in Contemporary Criticism*. New York, St Martin's Press.

Smith, W. (1964) Review of Huxley, A., 1964, *New Republic*, 150, p. 28.

Snyder, A. (1926) 'Blasphemer's Plateau', *Amazing Stories*, October, pp. 556–68.

Snyder, C. (1902) 'Newest conceptions of life', *Harper's Monthly Magazine*, 105, p. 856.

Snyder, C. (1912) 'Theory of life', *New York Times*, 6 October.

Sonneborn, T. (ed.) (1965) *The Control of Human Heredity and Evolution*. Macmillan.

Sonneborn, T. (1968) 'H.J. Müller, crusader for human betterment', *Science*, 162, pp. 772–6.

Squier, S. (1994) *Babies in Bottles: Twentieth-Century Visions of Reproductive Technology*. New Brunswick, NJ, Rutgers University Press.

Stableford, B. (1979) 'The utopian dream revisited', *Foundation*, 16, pp. 31–54.

Stableford, B. (1991) *Sexual Chemistry: Sardonic Tales of the Genetic Revolution*. Simon & Schuster.

Stableford, B. (1995) '*Frankenstein* and the origins of science fiction', pp. 58–74 in David Seed (ed.) *Anticipations: Essays on Early Science Fiction and its Precursors*. Liverpool, Liverpool University Press.

Stahl, W. (1995) 'Venerating the black box: magic in media discourse on technology', *Science, Technology and Human Values*, 20, pp. 234–58.

Steiner, G. (1968) 'It happened tomorrow' *New Yorker*, 44, p. 257.

Steiner, G. (1978) 'Has truth a future?', *The Listener*, 12 January, pp. 42–6.

Stevenson, E. (1929) 'What is life?', *Scientific American*, 140, pp. 18–19.

Stockridge, F. (1912) 'Creating life in the laboratory', *Cosmopolitan*, 52, pp. 774–81.

Stone, A. (1994) 'Fictions of the age of anxiety, 1945–1963', Chapter 2, pp. 33–62 in *Literary Aftershocks: American Writers, Readers and the Bomb*. New York, Twayne Publishers.

Storr, A. (1968) 'The future of the human body', *Sunday Times*, 21 April, p. 57.

Strehl, R. (1955) *The Robots are Among Us*. New York, Arco.

Sussman, H. (1968) *Victorians and the Machine: The Literary Response to Technology*. Cambridge, MA, Harvard University Press.

Taine, J. (1959) *Seeds of Life*. Panther Books.

Taviss, I. (1972) 'A survey of popular attitudes toward technology', *Technology and Culture*, 13, pp. 606–21.

Taylor, G. (1968) *The Biological Time Bomb*. Thames & Hudson.

Telotte, J. (1995) *Replications: A Robotic History of the Science Fiction Film*. Urbana, University of Illinois Press.

Thody, P. (1973) *Aldous Huxley*. Studio Vista.

Thornberg, M. (1987) *The Monster in the Mirror: Gender and the Sentimental/Gothic Myth in 'Frankenstein'*. Ann Arbor, MI, UMI Reseach Press.

Tobey, R. (1971) *The American Ideology of National Science, 1919–1930*. Pittsburgh, University of Pittsburgh Press.

Todd, D. (1995) *Imagining Monsters: Miscreations of the Self in 18th Century England*. Chicago, University of Chicago Press.

Toulmin, S. (1977) 'DNA and the public interest', *New York Times*, 2 March, p. 23.

Toulmin, S. and Goodfield, J. (1962) *The Architecture of Matter*. Hutchinson.

Toumey, C. (1992) 'The moral character of mad scientists: a cultural critique of science', *Science, Technology and Human Values*, 17, pp. 411–37.

Toumey, C. (1996) *Conjuring Science: Scientific Symbols and Cultural Meanings in American Life*. New Brunswick, NJ, Rutgers University Press.

Troland, L. (1917) 'Biological enigmas and the theory of enzyme action', *American Naturalist*, 51, pp. 321–51.

Tropp, M. (1990) *Images of Fear: How Horror Stories Helped Shape Modern Culture (1818–1918)*. Jefferson, NC, McFarland.

Tucker, A. (1966) 'Science today – a new breed of man?', *Guardian*, 22 February, p. 6.

Tucker, A. (1969) 'Conception in the lab', *Guardian*, 15 February.

Tucker, A. (1974) 'The life stylists', *Guardian*, 19 July, p. 13.

Tucker, A. (1978) 'The brave new world of test tube babies', *Guardian*, 27 July.

Tudor, A. (1989a) *Monsters and Mad Scientists: A Cultural History of the Horror Movie*. Oxford, Basil Blackwell.

Tudor, A. (1989b) 'Seeing the worst side of science', *Nature*, 340, 24 August, pp. 589–92.

Turner, G. (1978) *Beloved Son*. Sphere Books.

Turner, J. (1980) *Reckoning with the Beast: Animals, Pain and Humanity in the Victorian Mind*. Baltimore, MD, Johns Hopkins University Press.

Turney, J. (1994) 'In the grip of the monstrous myth', *Public Understanding of Science*, 3, pp. 225–31.

Turney, J. and Balmer, B. (in press) 'The genetic body', in J. Pickstone and R. Cooter (eds) *Medicine in the Twentieth Century*, Harwood.

Tyndall, J. (1878) 'Spontaneous generation', *Nineteenth Century*, 3, pp. 22–47.

Van Dyck, J. (1995) *Manufacturing Babies and Public Consent: Debating the New Reproductive Technologies*. Macmillan.

Vasbinder, S. (1984) *Scientific Attitudes in Mary Shelley's 'Frankenstein'*. Ann Arbor, MI, UMI Research Press.

Vaughan, P. (1970) *The Pill on Trial*. Weidenfeld & Nicolson.

Verrill, A. (1927) 'The Plague of the Living Dead', *Amazing Stories*, April.

Vyvyan, J. (1971) *The Dark Face of Science*. Michael Joseph.

Waddington, C. (1968) 'Finger on the biological button', *Science Journal*, May, pp. 114–16.

Waddington, C. (1969) 'A matter of life and death', *New York Review of Books*, 5 June, pp. 29–34.

Wade, N. (1974) 'Genetic Manipulation: Temporary Embargo Proposed on Research', *Science*, 185, p. 332.

Walter, G. (1950) 'An imitation of life', *Scientific American*, May, pp. 42–5.

Walter, G. (1956) *Further Outlook*. Duckworth.

Walters, L. (1975) *Bibliography of Bioethics*. Gale Research Co.

Warrick, P., Greenberg, M. and Olander, J. (1978) *Science Fiction: Contemporary Mythology*. New York, Harper & Row.

Waters, K. and Van Helden, A. (eds) (1992) *Julian Huxley: Biologist and Statesman of Science*. Houston, Rice University Press.

Watson, J. (1977) 'In defense of DNA', *New Republic*, 25 June, pp. 11–14.

Watson, J. and Tooze, J. (1981) *The DNA Story: A Documentary History of Gene Cloning*. New York, W.H. Freeman.

Watt, D. (ed.) (1975) *Aldous Huxley: The Critical Heritage.* Routledge & Kegan Paul.

Weart, S. (1987) 'Nuclear fear: a history and an experiment', Chapter 23, pp. 529–50 in H. Tristram Engelhardt, Jr and Arthur L. Caplan, *Scientific Controversies: Case Studies in the Resolution and Closure of Disputes in Science and Technology.* Cambridge University Press.

Weart, S. (1988) *Nuclear Fear: A History of Images.* Cambridge, MA, Harvard University Press.

Wells, H.G. (1895) 'The limits of individual plasticity', *Saturday Review*, 19 January.

Wells, H.G. (1914) 'Some possible discoveries', pp. 397–408 in *Social Forces in England and America.* New York, Harper & Row.

Wells, H.G. (1946) *The Island of Doctor Moreau.* Harmondsworth, Penguin; first published 1896.

Wells, H.G., Wells, G. and Huxley, J. (1930) *The Science of Life*, 3 vols. Amalgamated Press.

Werskey, P.G. (1971) 'Essay review – Haldane and Huxley: *The First Appraisals*', *Journal of the History of Biology*, 4, pp. 171–83.

Werskey P.G. (1979) *The Visible College.* Allen Lane.

West, R. (1932) 'Aldous Huxley on man's appalling future', *Daily Telegraph*, 5 February, p. 7. Reprinted in Watt, 1975, pp. 197–202.

Williams, H. (1900) 'Some unsolved scientific problems', *Harper's Monthly Magazine*, 100, pp. 774–83.

Williams, R. (1978) *Government and Technology, Units 3–4 of T361: Control of Technology.* Milton Keynes, Open University Press.

Wilmut, I., Schnieke, A., McWhir, J., Kind, A. and Campbell, K. (1997), 'Viable offspring derived from fetal and adult mammalian cells', *Nature*, 385, p, 810.

Wilson, E. (1923) 'The physical basis of life', *Science*, 57, pp. 278–86.

Wilson, E. (1928) *The Physical Basis of Life.* New Haven, CT, Yale University Press.

Winstanley, M. (1976) 'Who knows their DNA?', *New Society*, 22 April, p. 192.

Witkowski, J.A. (1979) 'Alexis Carrel and the mysticism of tissue culture', *Medical History*, 23, pp. 279–96.

Witkowski, J.A. (1980) 'Dr Carrel's immortal cells', *Medical History*, 24, pp. 129–42.

Wolpert, L. (1997) 'Under the microscope', *Independent on Sunday*, 23 February.

Wolpert, L. (1977) 'What's all the fuss about', *Independent on Sunday Review*, 23 March, p. 50.

Wolstenholme, G. (ed.) (1963) *Man and His Future – CIBA Foundation Symposium.* Boston, MA, Little, Brown.

Wolstenholme, G. (ed.) (1973) *Law and Ethics of AID and Embryo Transfer.* Amsterdam, Associated Scientific Publishers.

Wood, C. (1969) *Sex and Fertility.* Thames & Hudson.

Wood, C. (1970) 'Towards a test-tube baby era?', *Science Journal*, April.

Woodward, K. (1997), 'Today the Sheep . . .' *Newsweek*, 10 March, p. 48.

Wright, P. (1970) 'Progress towards "test-tube" baby', *The Times*, 24 February p. 1.

Wright, S. (1994) *Molecular Politics: Developing American and British Regulatory Policies for Genetic Engineering, 1972–1982.* Chicago, University of Chicago Press.

Yoxen, E. (1977) 'The social impact of molecular biology'. Unpublished PhD thesis, University of Cambridge.

Zebrowski, G. and Warrick, P. (1978) 'More than human? Androids, cyborgs and others', pp. 294–307 in Warrick et al., 1978.

Illustration Credits

Index